Collecting DOLLS HOUSES AND MINIATURES

The splendid oak panelled drawing room of the Abbey Grange house. Some 25½in (65cm) wide, it contains a fine collection of black and gold lacquered furniture and cabinets with glowing red decorations. It presents a striking contrast to the more conventionally furnished rooms.

Collecting DOLLS HOUSES AND MINIATURES

NORA EARNSHAW

New Cavendish Books
London

Pincushion Press
Tampa

DEDICATION

This book is dedicated to my long suffering husband Edward, without whose help and support this book would never have reached fruition.

My photographer, Frank Newbould, deserves special mention for his patience; always unruffled, he worked long and hard under what were, at times awkward conditions.

photograph: Frank Newbould

First published in hardback 1989
by William Collins Son & Co Ltd
London - Glasgow - Sydney - Auckland -
Toronto - Johannesburg
Paperback edition first published in 1993
by New Cavendish Books

British Library
Cataloguing in publication Data
Earnshaw, Nora
Collecting dolls' houses and miniatures.
1. Dolls' houses. collecting
688.7'23'075

New Cavendish Books Ltd.
3 Denbigh Road
London W11 2SJ

ISBN 1 872727 86 7 (UK edition)

Printed and bound in Spain
Produced by Imago Publishing

Published in the USA by
Pincushion Press
5245 Baywater Drive
Tampa, FL 33615

ISBN 1 883685 03 6 (USA edition)

Date	***c. 1883***
Maker	***Unknown English***
Heigh	***43in (109cm)***
Width	***42in (107cm)***
Depth	***30in (76cm)***

This imposing wooden dolls house is of an earlier style than the date, 1883, painted above the entrance. The initials MC are also shown, but sadly, the provenance of this house is unknown. The house stands in the entrance of the Abbey House Museum, where it is known, because of the simulated pargetting, as the black and white *house*. The interior, not shown, is also of an early period and is, because suitable furniture difficult to find, only scantily furnished.

Courtesy: Abbey House Museum, Kirkstall, Leeds;

CONTENTS

FOREWORD

I hope that the illustrations presented in this book will both please and excite those who are beginning (or simply contemplating) a collection of dolls' houses and also entertain and interest established collectors. Beginners will, I trust, find much here to widen their interests and add to their knowledge and experts find much that is new to them – some rare items are shown here that have never before appeared in any book on dolls' houses.

Collecting dolls' houses is an addictive and stimulating hobby and one that can give great pleasure and satisfaction to collectors of all ages and persuasions. The hobby will inevitably lead to a peripheral study of architecture as a means of identifying and as an aid in evaluating acquisitions, and a working knowledge of architectural terms is essential. For this reason, I have extended the glossary to include some terms that are not mentioned in the text. A basic understanding of architectural trends is necessary to provide appropriate furniture, furnishings, fittings and embellishments for the houses you acquire – whether they are basic items or decorative ones and whether they are for the kitchen or for the attic – and last, but far from least, dolls' house dolls. All these and many more require study, and for this reason I strongly advise you to use a magnifying glass when you scrutinize the illustrations of interiors in this and other books.

The sensitiveness of my photographer, Frank Newbould, is revealed in all the illustrations he has contributed to the book. He is a consummate artist in, particularly, his use of lighting and he has devoted unlimited patience to achieving his aim of bringing every picture to vibrant life.

If you buy with care and without the expectation of profit or for investment, you will find there is much happiness to be gained from the absorbing hobby of collecting dolls' houses and miniatures and in sharing your interest with like-minded collectors. Enjoy your collection, be it of one house or of many.

Nora Earnshaw

Date **Mid-18th century**
Maker **Unknown English**
Height **85in (213cm)**
Width **53in (135cm)**
Depth **24in (61cm)**

This unusual and extremely rare painted pinewood house, which is in the style of the 1720s, has a façade with sash windows opening to four rooms, three with panelled walls and marbled fireplaces. The kitchen has a fitted dresser, a fireplace and steel spit engine. The dining room has Chinese wallpaper and a contemporary alcove, and there is a needlework carpet in the parlour. The four-poster bed in the bedroom has contemporary Dutch hangings, but the stand on which the house rests is later.

Formerly at Swarcliffe Hall, Birstwith, Harrogate, and in the tradition of the Greenwood family at Swarcliffe Hall, this fine, large house was once owned by the Sidgwick family, who lived near Skipton. Charlotte Brontë was the children's governess in 1839, but she was unhappy with the family and is known to have described the children as "little devils incarnate".

Courtesy: Roger Warner; photograph: Sotheby's, London

INTRODUCTION

To define the collecting of dolls' houses in a clear, coherent fashion is far from easy: the spectrum is vast, covering as it does baby houses, cabinet houses of great magnificence and room settings. There are the furniture, furnishings and inhabitants; there are houses made by estate carpenters, cabinet makers and model makers; there are home-made abodes, and, of course, the wide range of commercially manufactured houses, rooms and shops, made principally in Germany, Britain, France and in the United States towards the end of the 19th century and well into the 20th century.

I have endeavoured to collate information about the practicalities of collecting in a format that meets the needs of established collectors and that will also serve the basic requirements of aspiring collectors. I hope I have succeeded in my efforts to steer newcomers to the hobby in a direction that will enable them to gain the necessary experience and the confidence to make the right decision when a new purchase is contemplated. The best advice I can give is "to keep one's head and not to act on impulse"; buy what appeals to you, what you can accommodate and what you can afford.

I am not alone in thinking that there are many, many dolls' houses still to be discovered. They may be lying folorn and forgotten in barns, outbuildings of country houses or attics, or even be tucked away in a box-room, simply because the present owner may think "no one will be interested in Great Aunt Agatha's old plaything". If you are a true dolls' house aficionado there is always the urge to seek and find. Never give up hope. If you are meant to have a certain house, set of furniture or whatever, *it* will find *you*. After I started to write this book, a number of items appeared as if by magic from two minor dolls' houses. Some I went in search of, but others were "added" items in auctions. I was thrilled when I espied in a country auction the set of furniture illustrated on page 73.

Remember Addison's homily: "If you wish success in life, make perseverance your bosom friend, experience your wise counsellor, caution your elder brother and hope your guardian genius." I would add to the above advice, observe and be vigilant. On the subject of observation, I again urge you to have a magnifying glass close to hand while you are studying illustrations, particularly if the full interior of a large house is shown. You will be amazed how much more detail will emerge from wall and floor coverings and how much small ornamentation will be apparent on furniture. What may appear to be "blobs" to the naked eye will be revealed as items of food on tiny plates or miniscule objects in a nursery.

An English child playing with a dolls' house; a plate from a book published between 1900 and 1925.
Courtesy: Abbey House Museum, Kirkstall, Leeds; photograph: Frank Newbould

Regrettably, some of the faces of dolls' house dolls will appear to be less attractive. They were often painted by children working for a pittance, and yet some of the dolls' attire will spring to life, fabrics glow and tiny stitches show the attention to detail paid by seamstresses of long ago. Fabric designs and, most important, the hair styles of the dolls will benefit from enlargement and help you to work out the approximate date.

You may find too much detail given in text and captions about some of the items illustrated, others may be too brief – this will depend on your own preferences and interests. I have endeavoured to steer a middle course. In my opinion, many dolls' houses speak for themselves. The more you scrutinize the illustrations, the more you will observe, but if you merely glance at the illustrations without really studying them, skip the introduction and pay scant attention to lengthy captions and much of the text, you are, to a certain extent, not taking advantage of what you have paid for. Established collectors read a book from cover to cover without missing a word because they know there is so much information still to come to light and they hope to find additional knowledge.

This book will guide you in your search for knowledge. But, like all collectors, you need to make yourself as expert as you can in your chosen subject. Take every opportunity to look at and learn from exhibits in museums, displays at miniaturists' fairs, and houses and furniture offered for sale in specialist shops or at auction. Read other books, and magazines devoted to dolls' houses and miniatures. Only in this way will you acquire the experience that will enable you to judge and evaluate the items you come across in your search.

The interior of Kosy Kot (see page 125). In addition to the large advertisement for Pears soap on the inside of the façade, both rooms in this two-storey house have portions of Pears advertisements stuck on the walls. The metal washstand on the right has two basins and a tablet of soap with the name "Pears" impressed on it. The washstand does not belong in Kosy Kot, but the German wall telephone was moved out of the house to be included in the illustration; it bears the words: "Telefon 946 Ges. Gesch."

Also in the foreground are two styles of metal fireplace; these were removed from the two blue-roof houses shown on pages 124–5.

Courtesy: Abbey House Museum, Kirkstall, Leeds; photograph: Frank Newbould

1
BABY AND CABINET HOUSES
17TH TO EARLY 19TH CENTURY

Date 1735–c.40
Maker Unknown English
Height 61in (155cm)
Width 67½in (171cm)
Depth 26½in (67cm)

The illustrations of the Nostell Priory baby house seen opposite and on pages 18–19 and 84–5 show this unique house as it was in July and August 1988.

The house has nine rooms on three floors. Each room is different, but all have skirting boards, chair rails and cornices and reflect the architectural designs of the period.

This house is an object lesson in how to present a top-quality, 18th-century baby house: the uncluttered rooms, the polished wooden floors, the few carpets, the polished, uncarpeted stairs, the oak wainscots and the oak-panelled rooms are a delight. The rooms are painted in subdued colours, decorated with hand-blocked (albeit oversized) wallpaper or, as in the yellow drawing room, with scenes or *découpage* on suitably coloured background. The soft furnishings harmonize with each other.

One could not hope to emulate the magnificent fireplaces in this baby house; they are possibly the finest to be seen in any baby house of this period. Paintings by unknown hands appear throughout the house – above fireplaces and over doors in the dining hall, dining room and red bedchamber. This last, a room of great importance, has pediments above the doors, as do the study and yellow drawing room. The brass door furniture and escutcheons on the various doors glow brightly, and the locks work and the knobs turn.

Courtesy: Lord St Oswald and the National Trust; photograph: Frank Newbould

In 1772 Horace Walpole made a present of shells to Lady Anne Fitzpatrick when she was five years of age. He wrote the following few lines to accompany the gift: "May these gay shells find grace and room both in your baby house and sight".

In the 16th, 17th, 18th and even the early 19th centuries, baby houses, as they were then known, ranged from magnificent cabinets, made of oak for adult amusement only, to painted and rusticated miniature mansions and country houses; there were also room settings and lavishly equipped German and Dutch kitchens, designed to educate and instruct young girls in the art of housewifery, and even shops of all types, many made for instructional purposes also.

The earliest known English baby house dates from about 1675. It surfaced in 1988 and realized a five figure sum at auction. Unfortunately no provenance was given, and the house was devoid of furnishings (see page 100).

The earliest fully furnished English baby house dates, probably, from the last decade of the 17th century. This is the Ann Sharp baby house. Its young original owner, from whom it takes its name, was the daughter of John Sharp, Archbishop of York, and the god-daughter of Princess Anne (later to become Queen Anne). The house was a present from god-mother to god-daughter. It is a large cabinet house, nearly 6ft (2m) high, 5ft 6in (1.7m) wide and 1ft 6in (46cm) deep, fronted by two glazed doors, which, even when closed, permitted a full view of the interior, with its nine rooms and long top shelf running the width of the house. The house has been preserved almost as Ann left it; it is now in the possession of the Bulwer-Young family.

Interest in baby houses grew in Britain as the 18th century went on. The Edmond Joy wardrobe house (see pages 14–15) is an early example. A similar house by Edmund Joy may be seen in the Bethnal Green Museum of Childhood, London. As the interest in baby houses spread to Britain from continental Europe, so too did a fascination with miniature items, made of pottery, porcelain or silver, and perhaps originally intended as collectables for adults, but later made especially to furnish the baby houses. Dolls, too, were created in increasing numbers and in increasing variety.

An outstanding baby house of the first half of the 18th century was the Nostell Priory house (see pages 11, 18, 19, 84 and 85). It was first planned about 1735 and completed between that date and 1739–40. It is one of

England's finest extant baby houses. Nostell Priory is a mansion built, c.1713, in the Palladian style for Sir Rowland Winn, the Fourth Baronet, to an earlier design reputedly by the Yorkshire architect Colonel James Moyser. The design was executed and almost certainly modified by the rising young architect James Paine, who is known to have worked on the house from 1736 onwards. On the death of the Fourth Baronet in 1765, his son, also named Rowland, assumed the title and commissioned Robert Adam to complete the work on the house and a new wing. Thomas Chippendale, who was born in 1718 in Otley some 30 miles from Nostell, is believed to have worked at Nostell for the Fourth Baronet, and Chippendale certainly made many of the exquisite items of furniture in the Priory in his London workrooms for the Fifth Baronet. There is a long-held Winn family tradition that James Paine designed the Nostell Priory baby house and that Thomas Chippendale was responsible for much of the furniture therein.

Lady Winn, wife of the Fourth Baronet, and her sister Miss Henshaw took overall responsibility for the soft furnishings, and made most of them with their own hands. Imagine the long hours spent in the discussions – which fabrics to use from a choice of velvets, silks and chintz, because much of the materials may have been remnants from fabrics used in the main house. The two ladies had their own ideas, too, about the type of furniture they desired for the exquisite miniature abode. One wonders if they travelled to London in search of some of the miniature silver items. It is not known if they obtained the dolls in England or if they were commissioned and made on the continent; they may have been dressed by Lady Winn and her sister.

In 1953 Nostell Priory was given to the National Trust, but still in occupation is a direct descendant of Sir Rowland Winn, the Lord St Oswald. I gratefully acknowledge the permission given by Lord St Oswald and Roger Whitworth of the National Trust to attend at Nostell with my photographer, Frank Newbould, to give a full, in-depth pictorial account of the awe-inspiring Nostell Priory baby house and its magnificent contents. Words alone cannot convey its magic. Feast your eyes on the illustrations photographed by Frank Newbould. Under his sympathetic lighting the rooms glowed softly. He, too, fell under the spell of this masterpiece and agreed with me that Nostell Priory baby house has an "aura". Our one disappointment was that the façade was stored, for safety, in another part of the estate and was not available for photography. A description must, therefore, suffice. The exterior purports to bear a slight resemblance to the main house. The baby house rests on a stand, which is of a later date. There are fourteen windows at the front, and six sash windows at each side. There is an ingenious method of opening the house, not by hinged doors but by sliding the doors back in special grooves. That part of the frontage containing the panelled door and eight windows slides to the left, the remaining part of the façade slides to the right. The front is enhanced by four columns in the Ionic style.

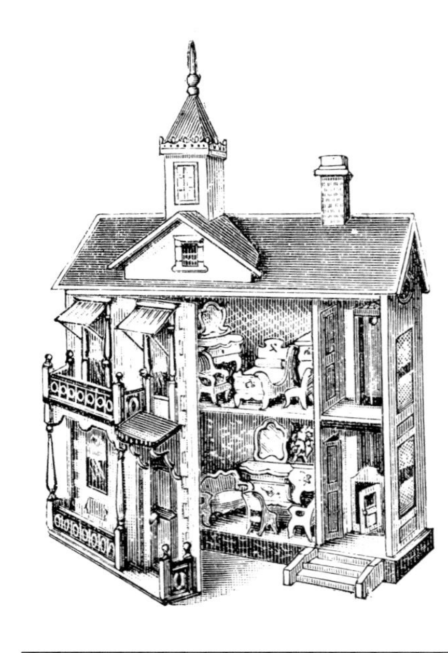

Illustrated on pages 22 and 23 is Quantock house. To say that this is one of my favourite baby houses is an understatement. It is a house that repays the closest study.

The Dower House, illustrated on pages 26 and 27, is a truly "lived-in" house, in which Vivien Greene has demonstrated to the full the art of bringing a dolls' house to life. In the downstairs parlour Grandmamma is sitting on the sofa nursing her youngest grandchild while her two grandsons gaze in dismay at the fallen tea cup – fortunately, nothing is broken. Mamma is bending down to retrieve the crockery while, oblivious of the mishap, her daughter is contemplating her piano practice. There are two handsome candle sconces on the walls on either side of the fireplace, the mantelpiece of which bears a rare mantel clock flanked by a pair of "milk glass" vases. A fire glows brightly in the grate.

When, many years ago, I first saw the magnificent and extremely rare

Date **Mid-18th century**
Maker **Unknown Dutch (with later German additions)**
Height **45in (114cm) (excluding stand)**
Width **65in (165cm)**
Depth **24in (61cm)**

The main structure of the fascinating Gontard cabinet baby house is 18th-century Dutch, but the pediment and lacquered stand were added during the 19th century in Germany. Note the two linen presses on the landing – the larger one was for bed linen, the smaller for handkerchiefs and serviettes. The two dolls in the hall date from the 19th century and are 7½in (18cm) tall. The house is discussed on page 25.

Courtesy: Historisches Museum, Frankfurt am Main

baby house that is illustrated on page 6, I completely lost my heart to it. It was being offered for auction, and I returned for three days before the auction to guard it from "non-dolls' house people", who were fiddling with the delicate sash windows and treating what I thought was going to be *my* house much too roughly for my liking. At the auction I carried on bidding to three times my limit, but was the under-bidder to Roger Warner. He is a worthy owner who loves the house as much as I do, and he has treated it handsomely and with the reverence it so richly deserves. I am indebted to him for allowing me to include the illustrations of this unique house in this book.

Mr Warner has furnished the house with exquisite taste. The imposing four-poster bed, the wall hanging, the magnificent bureau, the chandeliers, the candlesticks, the chopping block in the kitchen, the miniatures in the alcove in the dining room, the corner fireplaces in the two upper rooms – all these are right for the period and for the proportions of the rooms. I first saw this house in a rather forlorn state and with scarcely a stick of furniture, in need of much loving care and attention. I feel I can say: "Well done, Mr Warner. You have turned a relic into a miniature stately home."

The Castle Museum in York is always worth a visit, especially for dolls' house collectors eager to see the famous Yarburgh baby house, which was presented to the museum by its last owners, Lord and Lady Deramore.

Date **c.1709**
Maker **Edmund Joy**
Height **65in (165cm)**
Width **57in (145cm)**
Depth **26in (66cm)**

This rare and important William and Mary decorated child's wardrobe is in the form of a country mansion in the Anglo-Dutch style. The recessed broken outline is surmounted by a pierced spindle gallery, and the façade is composed of shaped gables with scroll ornament, the central gable containing a rectangular mirror. The three panel doors are inset with glazed windows, and there is a central portico and painted chequered steps. The doors have contemporary key plates and drop handles. The interiór is fitted with shelves on the left-hand side and a bank of eight drawers on the right-hand side. The drawers are painted with flowers. The house bears the signature of Edmund Joy and the date 1709 on the side. A similar house by Edmund Joy may be seen in the Bethnal Green Museum, London.

Courtesy: Phillips, London

This 18th-century house, which possibly dates from 1710–15, was undoubtedly designed for children's pleasure rather than as adults' treasure, a conclusion I reached after examining the façade. It is serviceable, but does not have the architectural details and elegance an adult would have demanded.

James and Ann Yarburgh lived at Heslington Hall and had twelve children. It is not known whether the eldest child, Henrietta, who was born in 1695, paid much attention to the baby house before her marriage in 1719 to the architect Sir John Vanbrugh. They met when Vanbrugh visited Heslington Hall during the building of Castle Howard for the Earl of Carlisle. The marriage adds interest to the provenance of this interesting house, which remained at the family home of Heslington Hall until its presentation to the museum.

Yarburgh baby house is a flat-roofed structure, 5ft 4in (1.6m) high (including an 11in (28cm) stand), 4ft 8in (1.4m) wide and 2ft (61cm) deep. The façade is painted an indeterminate shade of pale grey and the window mouldings and arched door casing are painted white. There are eight windows, and the arched door is glazed with nine panes topped by three fan-shaped panes, which form the arch. The interior is reached by nine separate winged doors, and there are nine rooms, three storeys and three bays. The house retains most of its original wallpapers, and some original furniture together with later additions. The kitchen appears to be little changed and still has a clockwork jack. One notable feature of the interior of the house is the clever impression that is given of internal doors – moulded architraves surrounding flat, painted marbling representing six interconnecting doors.

Every author is entitled to a favourite house, and of the ones on public view Nostell Priory baby house is my favourite, but Uppark baby house runs Nostell a close second. This miniature mansion is illustrated in many of the books listed in the Bibliography, but for novice collectors I feel I would be failing in my duty if I did not include a written description. Uppark baby house was built c.1730 for Sarah, the only daughter of Christopher Lethieullier, of Belmont, Middlesex. (The family had its origins in Belgium.) Sarah married Sir Matthew Fetherstonhaugh, and in February 1747 Sir Matthew and his young bride came to live at Uppark House, in the village of South Harting, near Petersfield. Sarah brought the baby house with her. It was evidently one of her most prized possessions and must have been a valuable toy even in those days; it may, indeed, have formed part of her dowry. As an only daughter, Sarah must have spent many long hours engrossed in the delights of her magnificent baby house, but one wonders how – maybe she perched on top of specially designed steps or even library steps, surely the only way in which she could have played with the house. The baby house stands some 7ft (2.1m) high, supported on an arcade of seven arches. The arcade is rusticated, as is the façade of the first storey of this three-storey three-bayed house. There are twenty windows, each containing twelve panes of glass. The arched door is unpretentious, but not so the other architectural details, which were obviously designed by an aficionado of the Palladian school. The centre bay has two rusticated columns and Ionic and Corinthian pilasters on the storeys above. The balustraded roof is dominated by seven statues – the purist would probably say they are over-large and rather ornate. I leave to the last the description of the pediment surmounting the centre bay: the tympanum bears a handsome shield painted with the arms of the Lethieullier family, and the whole is surrounded by intricately executed repoussé floral scroll-work. The house has been re-painted several times during its lifetime and is now a soft shade of grey.

The interior of the house is reached via nine "winged" openings. The furniture and furnishings are of the period; the dolls of wood and wax; and there are many silver items and examples of early English pottery and porcelain.

Two other architectural gems that are essential viewing are the Blackett house and the Tate house. The Blackett house, which is now at the London Museum, dates from c.1740. The façade is exceptionally finely detailed, and the interior does not disappoint. The house contains good furniture, some of which is original, although there have been later additions over the years. Surprisingly, the house retains its original wax dolls. The kitchen appears to be little altered; the outstanding mechanical jack is still in position and the spit rack is in place in front of the fire. This is a much publicized house, as is the Tate house, an impressive abode with superb architectural detailing to the façade and exceptional carpentry. Dating from c.1760, it was based on a Dorset country house of the period, but in the 19th century it suffered much additional attention and refurbishment. It is now splendidly displayed at the Bethnal Green Museum of Childhood in London. This museum contains more than thirty other houses, toys and dolls of many periods up to and including the present day, and it is always worth a visit.

If we turn now to the rest of Europe, we will not concern ourselves with relics in terracotta, wood, bone, stone, bronze, lead, remains of miniature rooms, funerary models and tomb relics. They may be seen in the Tarquinia Museum in Italy and the Metropolitan Museum of Art, New York. Theories abound regarding Greek grave relics and models. Karl Gröber expounds on dolls' houses: "They must have existed; for where else could use have been found for the little pieces of bronze furniture which have survived if not in a doll's room; while the many little utensils belong of a certainty to the furnishing of a doll's kitchen." Gröber also refers to, and quotes from, the reference of the Greek traveller Pausanias to a miniature ivory bed presumed to have been a toy of Hippodameia, the wife of Pelops. She had offered this toy to the Greek goddess Hera. Gröber comments "such doll furniture could be very costly". We may as well remain with Gröber and his preoccupation with dolls' houses or miniature rooms when he quotes: "If Plato can write that the future architect should from his earliest years busy himself with the building of houses, we can safely conclude that even in those days boxes of bricks, whatever they may have looked like, cannot have been wanting." But

OVERLEAF:
The kitchen of the Nostell Priory baby house (see page 11) is narrow but airy, with the walls painted to resemble stone. The cooking range occupies the back wall, on which we see the wooden spit rack and the steel jack. The dresser, which is 11in (28cm) high and wide, bears a lidded tankard. Close to the range is a varnished wooden plate rack with many silver plates. There used to be a silver tripod plate holder in the kitchen, but when I commented on its absence I was assured that the kitchen had always been as it is shown here. When I was measuring the doll in the red bedroom, I noticed the silver plate rack, upside-down behind a silver wash-stand. Even though I pointed this out to the staff, I was told that there was an inventory of the items in each room, and that "this has always been here". In the kitchen, a splendid array of silver items rests on the polished table, there is a silver chamberstick on top of the chopping block, and on a stand near to the door is a large spirit kettle. The door is panelled in dark wood and has clean-cut architraves and a brass knob. The immaculately attired chef and his colleague (the footman in the dining hall) are carved in wood, the boots carved in one with the well-shaped legs. The features are painted and, a thoughtful touch, one hand on each figure has a space to enable the model to grip a piece of equipment. An ivory mouse in the foreground is followed by an outsize black dog, the size of a calf.
Courtesy: Lord St Oswald and the National Trust; photograph: Frank Newbould

The dining hall (above) of the Nostell Priory baby house has a magnificent balustraded staircase, with a curtail and small gallery. Note the nine-branched chandelier, the paintings, the rare cabinet-made furniture, the longcase clock (11in (28cm) high), the silver on the table, the polished wooden floor and the footman (see page 17). The two chinoiserie cabinets (left) in the yellow drawing room are decorated with woodland and semi-classical scenes, painted and lacquered over gesso. The cabinet on the left opens to reveal six drawers; it is 7¼in (18.5cm) high, 4½in (11cm) wide and 2in (5cm) deep. The chest on the right is 4½in (11cm) high and wide and 2½in (6cm) deep. Note especially the quality of the stands: from a distance they almost resemble tortoiseshell.

Courtesy: Lord St Oswald and the National Trust; photographs: Frank Newbould

children have always tried to emulate adults – boys have played with swords and girls must always have played at keeping house, so it is not unreasonable to suppose that toy houses or miniature rooms have been among children's playthings since ancient times.

We are on firmer ground when we move on to the earliest European dolls' houses. These were cupboard or cabinet houses – so called because they were rooms displayed in a cabinet rather than in a true model of a house. (The name "dolls' house" came into use only in the first half of the 19th century – before that the term used was baby house.) We know that wooden toys and miniature furniture as well as room settings were made in Nuremberg and other parts of Germany in the early 15th century, but the earliest cabinet house of which we have any detailed record is a four-storey baby house (*Dockenhaus*) commissioned in 1558 by Duke Albrecht V of Bavaria. Although apparently intended for his daughter, it was placed by the duke in his art collection – perhaps because he thought it too fine and costly to be used as a plaything. It is believed to have been destroyed in 1674, when the palace in Munich in which it was housed was gutted by fire. We owe our knowledge of it to a detailed inventory of the house and its contents that was drawn up by the duke's chamberlain, Johann Baptist Fickler, in 1598.

I cannot say too often that this book is primarily about collecting dolls' houses, and throughout, my intention has been to cater for both novice and established collectors. Illustrations of, and details about, German *Dockenhauser* (baby houses) have appeared so often in previous books on baby houses that dolls' house collectors of long standing know them off by heart. For this reason I have included only brief details of these magnificent, essentially German creations, and I strongly urge both beginner and established collectors alike to visit the Germanisches National Museum in Nuremberg, where they will be able to admire several unique 17th-century *Dockenhauser*. The earliest is a three-storeyed, south German house of 1611. It is 8ft (2.4m) high, 6ft 5in (2.0m) wide and 2ft (61cm) deep. The hall is part papered with tapestry-like murals whose designs can be traced back to an engraving by Jan Selder die Altere and are taken in part from a drawing by Dirck Barendsz (1534–92). The house has stairs, arched landings and a well-equipped kitchen. Another house (known as the Stromer house), of about 1639, is, although smaller, even more lavishly fitted out. The Kress house, dating from the mid-17th century, is a delight, with its three splendid staircases, a profusion of contemporary plenishments, interesting nursery, well-turned balustrading and twin gables – altogether a most complicated abode for family and servants.

More often than not 17th- and 18th-century *Dockenhauser* were equipped with locally made miniature items, the work of craftsmen working within the confines of the guilds – copper utensils made by a coppersmith; tiles and other ceramics by a potter; all tin items, from kitchen and culinary requisites to kettles and pewter tableware, by a tinsmith, wooden furniture, cabinets and so on by a cabinetmaker.

German baby houses often contain a merchant's shop and a separate room for domestic stores. A variety of containers not seen in Britain but peculiar to different regions of Germany and other European countries was needed to hold live animals, which were often fattened up in the kitchen with scraps left from the table – for example, cages for poultry or containers to hold live fish. Sleeping quarters were provided for servants in the kitchens. The *Dockenhauser* provide us with this glimpse into the past and enlarge our knowledge beyond the mere confines of our main interest, collecting dolls' houses and miniatures.

An interesting Nuremberg dolls' house, of about 1673, containing an elaborate kitchen is now in the Bethnal Green Museum of Childhood, London. There is also a cabinet housing an elaborately fitted kitchen, believed to be Dutch. One of the best examples of Nuremberg kitchens to be seen in Britain today is in the Bowes Museum, Barnard Castle, Co. Durham; it is shown on page 53.

Dutch cabinet houses are as interesting and as well documented as the German. They contain, too, far more silver miniatures than do the Nuremberg houses. For centuries Holland has been famed for its silversmiths' skills in producing these silver miniatures of household objects, utensils and figures. Although individual craftsmen in England and in Germany made similar, and no less beautiful, miniatures, they could not match the Dutch for the sheer quantity of their production.

"Petronella" was a favourite girl's Christian name in the 17th-century Netherlands, and three superb cabinet houses are connected with the name. Fortunately, all three are documented and well-preserved for us to see today, with their contents – miniatures, paintings and specially commissioned furniture and furnishings. The house of Petronella Dunois (married name, Willem), in the Rijksmuseum, Amsterdam, is an eight-roomed cabinet house from the last quarter of the 17th century. It is inhabited by a dozen or so wax dolls that are beautifully costumed and detailed – the mistress of the house has a pearl necklet at her throat and a bracelet on her wrist. Her long hair is well arranged, and there is gold-lace trimming on her petticoat, which is topped by a silken open robe. In the same museum and not to be missed, is the Petronella Oortman house of about the same date. (Petronella Oortman married Johannes Brandt in 1686.) The museum also has a painting of this house by a contemporary artist, Jacob Appel (1680–1751). The nine-roomed interior, with its landscape paintings and fine furniture, is exquisite. The exterior is covered in tortoiseshell veneer, so expense must have been of no consequence. The painting shows the house peopled with dolls, but there are none in the cabinet now on view in the museum. The lack gives the house an empty, static feel. The third Petronella house, by contrast, is full of bustle and activity, with doll occupants in almost all the 11 rooms. This is the Petronella de la Court walnut-cased cabinet house in the Centraal Museum, Utrecht. (She married Adam Oortman in 1649.)

I shall mention here only one Dutch house of the 18th century. It is the Sara Ploos van Amstel house in the Gemeentemuseum, The Hague.

The exterior of the mid-18th-century English dolls' house whose interior is illustrated and described on page 6.
Courtesy: Roger Warner; photograph: Southeby's, London

Date **1700–25**
Maker **Unknown English**
Height **64in (163cm)**
Width **47in (119cm)**

Quantock, an architectural gem, is a baby house dating from the first quarter of the 18th century. It is in the classical style favoured by the first Lord Burlington, Colin Campbell, William Kent and Giacomo Leoni. (Vivien Greene detects a similarity to the work of John Wood Senior, who moved from London to Bath.) The façade glows with life. The oak structure is obviously the creation of a master craftsman, made with loving care and skilful attention to the carved details. Note especially the separate application of the panels indicating rustication, the columns with neat entablatures, the well-designed pediment and the unobtrusive finials. The house has 27 windows, glazed with the original crown glass, and the base contains the customary storeroom and game larder, both essential features of a great house. The interior of Quantock lacks only inhabitants to evoke further memories of the past. The hall floor is inlaid in dark and light squares. The staircase, redolent of past use and with a charm all its own, leads to an upper landing. The four large rooms have architectural detailing to the doors and original fine fireplaces. Furniture for a house of this period is extremely rare and, therefore, larger furniture of the period 1835–50 has been installed (the smaller furniture is contemporary with the age of the house). The house still possesses its original iron key – a rarity and great joy to its owner. Quantock surely ranks as one of the finest baby houses extant.

Courtesy: Mrs Graham Greene, The Rotunda Museum of Dolls' Houses, Oxford; photograph: Frank Newbould

"A VISIT TO A CITY OF TOYS" BY KATHLEEN SCHLESINGER, FROM *GIRLS' REALM* ANNUAL, 1902

We will wend our way to the oldest and most picturesque part of the town, on the Sebaldus side of the river Pegnitz. Here, on the side of the hill in the Brunnengasschen, the massive oak door at No. 14 stood ajar. I rang a bell, and at the head of an oak staircase blackened with age, winding and low, stood a figure in the gloom inviting me to step up into the counting house to see Herr Christian Hacker, a builder of dolls' houses, stables and shops, in whose work all English girls who have at any time loved their dolls will feel interested. Through the courtesy of his son, I was enabled to watch the growth of many dolls' homes. The rooms (where the houses were made) were all low, with raftered ceilings, the house itself was the residence of some of the patrician families of Nuremberg more than five hundred years ago.

In the machinery room, where the wood is sawn into proper lengths and cut ready for building the thousands of miniature dwellings which are exported every year to England and America, we only remained a short time, for the noise was deafening. We passed on to where the pieces were being fitted together. A smile lit up the face of the grave-looking workman when he found that I wished to photograph him. At a bench to his left, small fittings such as doors, staircases, chimney pieces, were growing under the deft fingers of a lad; another turned out stacks of chairs, tables, dressers and other furniture for the houses.

At the other end of the same workshop stood a large doll's house in the course of erection and a woman putting in the fixtures of a grocer's shop with her glue brush. There are fifty employed altogether in Herr Hacker's factory, which is a thoroughly typical one, and a third of these workers are women.

In the painting room, the women were working, putting in a white ground and a pale blue border, which they drew without a hand-rest with never a wavering or ragged edge. Before we could take a photograph, a barricade of smoothly planed shops waiting for their coats of paint had to be cleared; there were many such in the workshop dividing it into lanes, where the women sat at work.

After thus seeing all the various processes, I enjoyed a visit to the storerooms where the finished toys are kept prior to exportation. The dolls' houses are of two distinct styles, English and German, and I was charmed with two of the former. Herr Hacker, who has spent many years in London, had these made in exact facsimile of two houses in Clissold Park and at Balham. They are beautifully finished off in every detail; the window sashes are made to lift . . . the gates and doors to open, the bells to ring; inside is a winding staircase, on to which the doors of the rooms open.

Article on Christian Hacker, see also page 41.

This cabinet house, dating from about 1743, contains a collection of rare items, many of them already antiques when Sara purchased them – she bought three cupboards crammed with miniatures at an auction. The nine rooms are filled with beautiful treasures. This, incidentally, is undoubtedly one of the best documented of all Dutch houses – Sara kept all the bills and accounts relating to the house.

In the Historischemuseum, Frankfurt-am-Main, is the splendid Gontard cabinet house (see page 13). Each room is evocative of both Dutch and German influences, and this is not surprising in view of its past history and the number of hands through which it passed. Tradition has it that friends in Holland gave the cabinet and most of the contents to Susanne Maria, a daughter of the von Gontard family. Several generations later, it reached the hands of Mrs Belli-Gontard, and she it was who rearranged some of the rooms to resemble how she thought they would have looked several decades earlier. Items of German furniture were added over the years, and the whole combines to give a valuable glimpse of the past through the eyes of the Dutch and German craftsmen who contributed to this enchanting house. Note the linen press and small serviette press on the upstairs landing between the bedroom and kitchen. (The tradition is carried on to the present day: after visiting the museum I was amazed to see the modern equivalent, an electric roller ironer, on the wide landing of my hotel in Darmstadt.) The room to the left of the hall-cum-living room is the storeroom and larder, and the larger items would be stored behind the slatted double doors. The charming mural painted on the wall behind the gallery above the storeroom was known as a "vista", and this one may have been intended to give the impression of a view seen from a window.

This cabinet baby house bears some resemblance to the interior of the more lavish, earlier Petronella Dunois cabinet house in the Rijksmuseum, Amsterdam. I am very fond of this house and spent over an hour standing in front of it making notes.

In 17th-century France, individual room settings, some quite large, seem to have been preferred to full-sized cabinet or baby houses. In the first decade of the century, the Dauphin (later Louis XIII) and his sisters played with rooms and toys of various kinds. The journal of Jean Héroard, the Dauphin's physician, which records details of life in the nursery, tells us that "the little girls had several small rooms furnished with dolls' house furniture. One of these had a bed with Mamma and the baby in it, while the midwife stood by in attendance". The children also had "pewter plates and dishes of a size to play with at dolls' dinner parties" and some silver and earthenware sets, too. Elsewhere, it is recorded that in 1630 Cardinal Richelieu, Louis XIII's chief minister, gave a miniature room to the Princess d'Enghien, together with six dolls posed in a tableau of domestic life.

There are so many interesting 18th century houses to see in this country, on the Continent and other countries, but I hope that you enjoy the "appetizers" shown in this chapter.

LEFT:
Date **Late 18th century**
Maker **Unknown English**
Height **25in (64cm)**
Width **37in (94cm)**
The façade of the Dower House has been painted a mellow buff colour. The handsome oak front and side doors are panelled, and the house is well supplied with windows. Each of the four large rooms has an arched side window and two large front windows. The landing has one window. The overall impression is of a sturdy, well-used house of infinite charm and character that must have delighted several generations of children and that is continuing its life by enchanting its owner and other adult dolls' house collectors.

OPPOSITE ABOVE:
This elegant house, aptly named the Dower House, is typical of an abode built for a widowed mother in the grounds of a large estate and within calling distance of the family mansion. Notice especially the door leading to the kitchen at the side of the house, together with the arched windows. The top landing is unusual: it has a back door, possibly to give the impression of leading to the servants' quarters. The hall below boasts a splendid carpet on its well-constructed stairs. The walls of the four spacious rooms and of the hall are in the original paint, now faded. The elaborate base does not belong to the house.

ABOVE:
The kitchen of the Dower House is a working room, with everyone going about their appointed tasks. Note the flat iron resting on the fender in front of the Georgian fireplace, the laundry equipment, the food on the table in front of the side window, the notice board behind the side door, the copper kettle and stewpot, the fish moulds on the walls, the pewter kitchen utensils and the apron carelessly thrown on to the rocking chair.
Courtesy: Mrs Graham Greene, The Rotunda Museum of Dolls' Houses, Oxford; photographs: Frank Newbould

2
DOLLS' HOUSES
19TH AND 20TH CENTURIES

French children playing with a dolls' house: an engraving from 1875–1900.

RIGHT:
Date 19th century
Maker Unknown German
Height 27in (69cm) (to top of chimneys)
Width 29in (74cm)
This metal dolls' house has been painted to simulate stone. It originally had glazed, Georgian-style windows in the front, which opens to show four rooms, each with its original fire surround. This house is similar to the one described in the patent dated 29 June 1899, which was taken out by Bernard Scheer, a "tinman" of Burgstadt, Saxony, who devised a system of making dolls' houses from sheets of metal that slotted together.
Courtesy: Sotheby's, Chester

The selection of illustrations in this chapter shows baby houses of distinction at one end of the spectrum, commercially made, fine quality wood and lithographed dolls' houses down, at the other end, to the cheap, two-roomed houses played with by poorer children, who yet nevertheless applied their own personal touches to their treasured playthings. The two-roomed house, illustrated on page 9, with advertisements for Pears soap pasted on the walls both inside and out, is a humble house, but it conjures up delightful images of the child with her pot of paste; it is a house of character and charm, proving that a large cheque book is not always necessary to start a collection of dolls' houses. It is possible to get as much pleasure from looking at small as at elaborate houses, sometimes more, for the larger houses, although exquisite to look at, are often not so evocative of the past as children were not allowed to touch them. Queen Mary's dolls' house at Windsor is exquisite – a 20th-century masterpiece that is known all over the world; Titania's Palace, now at Legoland, Billund, Denmark, is another magnificent, equally well-known creation. Both excite many collectors of miniatures, but they are too perfect for the liking of other collectors. Understandably perhaps, children seem to identify with Titania's Palace rather than with Queen Mary's dolls' house.

The first quarter of the 19th century saw the gradual transition from baby houses to dolls' houses, from adults' treasures to children's educational toys. At this time began to appear room settings, dairies, kitchens, miniature schoolrooms, Noah's arks and, the game that brother and sister could enjoy together, playing at shops; village stores, building bricks and so on became increasingly widely available from about 1830, when more and more dolls' houses were manufactured, and the term "baby house"

FROM A PATENT, No. 13,511, DATED 29 JUNE 1899, TAKEN OUT BY BERNARD KARL EMIL SCHEER, A MASTER TINMAN, OF BURGSTADT, SAXONY, FOR IMPROVEMENTS IN TOY HOUSES

This invention for improvements in toy houses relates to small sheet metal houses used as toys and for similar purposes, and has for its object to provide means for readily putting the houses together entirely as may be wished, or required, or taking them apart again, which is of considerable advantage in relation to transport and packing.

The form and arrangement of the invention will be better understood with the assistance of the accompanying drawing.

faded from common use. Educational toys came to the fore as the years passed, and children were taught to learn as they played and encouraged to create toys from discarded boxes and scraps of material.

If we now turn to the mid-19th century, we find the discovery of the process of lithography, a boon to many, not least the makers of dolls' houses, toy farms, shops and Noah's arks and so forth. The manufacturers could now embellish their products, covering them with brightly coloured lithographed paper. The inventor of the process was Aloys Senefelder, who was born in Prague on 6 November 1771 and who died in Munich on 26 February 1834. Senefelder was originally a playwright, but he was too poor to pay for the necessary engravings and decided to try to produce them himself. In his first experiments he used Bavarian limestone as well as metal plates, but two years of experimenting led to the discovery of flat-surface printing, and in 1818 Senefelder filed his discovery in the Vollstandiges Lehrbuch der Steindruckerey.

When we reach the middle of the 19th century, we find increasing numbers of dolls' houses made for the children of the wealthy and middle classes. Poor children still had to be content with whatever could be fashioned from wood or cardboard, but this was a period in which goods were packed in stout wooden boxes to survive the journey from manufacturer or producer to their destination, whether it was grocer, ironmonger or tea merchant. Many dolls' houses were made from orange boxes or from soap boxes, for example, and these houses often survived and were improved upon. The backs of some modest dolls' houses still bear the name of the soap or candle manufacturer whose products were originally packed in the box. I once saw a three-storey dolls' house that had been made from a wooden packing case and that bore the legend "Waring & Gillow Ltd, Removers, London".

Date **c.1880**
Maker **Unknown English**
Height **34in (86cm)**
Width **40in (102cm)**

This substantial wooden dolls' house has two storeys. The façade has a central panelled door, with a window above, and symmetrical wings to either side with Georgian-style windows. The simulated red brick façade has cream window frames, mouldings and other woodwork. The flat-topped, pitched roof has a chimney at each end and a gable above the entrance. The front opens in three sections to reveal a plain wooden central stairway and half-panelled hall, three papered rooms with fireplaces and, on the lower floor, a fully panelled room. The interior doors are panelled.

Courtesy: Sotheby's, Chester

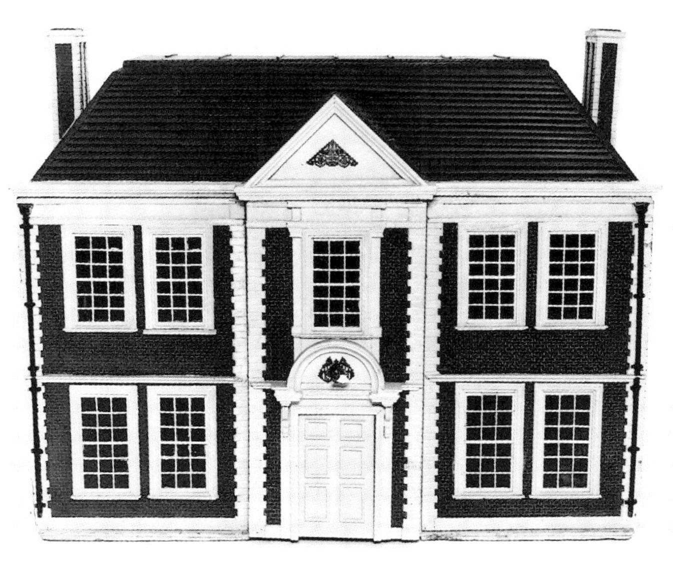

ABOVE:

Date **c.1889**
Maker **English**
Height **25in (64cm)**
Width **37¾in (96cm)**
Depth **20in (51cm)**

Regency villa was made by an eccentric English vicar who raised peacocks. The house has three storeys and nine rooms, including a conservatory and attached stable. The house has two towers, a balcony, a courtyard with a working bell on the fence and many styles of glazed windows. The wall and floor papers throughout are original. A twin of this house may be seen in the Bethnal Green Museum, London. The rear view of Regency villa is shown left.

Courtesy: Kansas City Toy and Miniature Museum

In this and other chapters of this book there are a number of patents. These may seem boring to novice collectors, but the patents are included, in as much detail as space will allow, for identification purposes. A large number of patents were taken out for so-called folding or collapsible dolls' houses and room settings, and for folding kitchens, shops and all manner of toys. Many patents were taken out during World War I, when manufacturers were urged to make up the domestic shortfall because of the lack of imports. The biggest surprise was to find dolls' houses not listed as such in patent abridgements but under the umbrella heading "boxes", but, of course, many dolls' houses are little more than boxes, possibly only two floors and two rooms – two boxes – or they may be built up into sixteen-roomed mansions. When I have managed to match extant items to the patent, the relevant items are illustrated. Research into Bruno Ulbricht led me to try to discover the British agent who acted for this company and that led me to William Seelig, a leading London agent, and I have listed on pages 32 and 33 the companies represented by Seelig.

The "Occupation Abstract" prepared by Henry Mayhew (1812–87) in the mid-19th century contains the following information. Mayhew calculated that there were some 400 toymakers and 150 toy merchants and dealers in London alone, plus approximately 1,300 scattered throughout the rest of the country. Some dealers were importers only, others flourished as both importers and exporters. The classifications used to describe traders in their advertisements and in the records of registered marks are, more often than not, insufficiently detailed for us to say with any certainty that any one specific dealer handled dolls' houses. The

BRUNO ULBRICHT,

Nuernberg (BAVARIA).

FACTORY OF FINE METAL TOYS.

A.B.C. Code. 5th Edition. Cable Word, "CLOCKENSPIEL."

Agent **W. SEELIG,**

23, White Street, Moorfields, London.

CREATOR of the following ORIGINAL Products :—

LEFT:
The rear interior of Regency villa. Note the large bell, adjacent to which, in the house itself, is an earth closet.
Courtesy: Kansas City Toy and Miniature Museum

classifications tended to be much more general, referring mostly to toys, either wooden or tin, which gave the merchants plenty of scope.

From the beginning of the 20th century the importation into Britain of dolls' houses, rooms, kitchens, shops, dolls' house dolls, furniture and, in fact, all toys, gained momentum, interrupted only by World War I. After 1918, trade between Britain and the rest of Europe started up again with renewed vigour. Germany, forced to pay reparation debts, was desperately short of currency and undercut British manufacturers, many of which had started up during the war to fill the gap left by the lack of German imports but which were now forced to close within a few years of the end of the war. Lines Brothers weathered the depression and went from strength to strength. Lines' houses, in particular those made before World War II, are often the first dolls' house acquired by new collectors, and most established collectors either still have or have had more than one example of the company's output in their collections.

Many manufacturers appointed agents at this time, while others registered trademarks or tradenames. Some of the main manufacturers and their marks and some agents are listed on pages 116–31, but it is only a representative selection of the many people and companies active

FROM A PATENT, No. 8581, DATED 15 APRIL 1903, TAKEN OUT BY KARL TRÄGER OF LEIPZIG-MÖCKERN FOR IMPROVEMENTS IN THE MANUFACTURE OF TOYS

This invention has reference to improvements in toys and more particularly to a doll's room or portion of a doll's house.

In order that it may be possible conveniently to pack a doll's chamber I propose to use as the floor of the room the folding box, which serves for packing up the doll's chamber and its model furniture, and on the floor so composed the walls can be set up as desired so as to form one or more chambers. . . . The floor of the doll's room consists of the box, which is intended for packing the parts of the chamber. The two parts of the box . . . are hinged or otherwise joined together and at such a level that, when laid adjoining each other they are suitable for forming the level floor of a doll's room. In this position they may be so united by means of a hook . . . that they are fastened and locked together and enable the doll's chamber to be built up and adjusted as to height or level.

The room or chamber is formed by placing a number of walls on this floor and connecting them together. These walls are preferably all made of the same size and so arranged that two adjacent walls may be joined together by hooks or bolts. The connection of the walls with the floor may be effected as desired, for example, by providing the walls with pins or dowels and the floor with corresponding holes or sockets, in which the dowels are inserted. . . .

A single doll's chamber may thus be formed as shown or by the employing of one or more intermediate or partition walls several such chambers. The walls may be put together so as to be of rectangular or polygonal form.

In order to pack the parts together it is only necessary to remove the walls and put them in the top of the box, the furniture, however, being placed in the bottom of the box. In this way the whole doll's chamber can be easily packed in one box.

A 19th-century metal money box (bank) made in the shape of a house and a 20th-century English-made tin money box in the shape of an electric oven; the latter is marked "Wright's Eureka Works, Birmingham/Made in England 555".
Courtesy: Abbey House Museum, Kirkstall, Leeds; photograph: Frank Newbould

in this area. It is, however, worth looking in some detail at the activities of William Seelig, an agent who was working in London in the early years of this century. Seelig represented at least 30 European companies, acting as agent for: Burger Haidenau of Berlin (manufacturer of tea sets and printed dolls' house wares and toys); Beck & Glaser of Königsee (manufacturer of dolls' house artefacts and tea sets from 1897 to about 1930); Carl Brandt of Gössnitz (manufacturer of wooden toys); Canzler & Hoffman of Berlin and Sonneberg (manufacturer of all types of dolls, from bisque to celluloid, including dolls' house dolls, from 1906 to at least 1930, which used *Caho* as a trademark and marked some goods or boxes D.R.G.M.); Cosmos Publishing Co. of Stuttgart (manufacturer of paper and card products and games); F.W. Gerlach of Nuremberg (manufacturer of gilt-metal and tin ware stoves, kitchens, culinary utensils and dolls' house furniture); J.D. Kestner Jr of Waltershausen (manufacturer of a wide variety of dolls including some dolls' house dolls, see page 111, from 1805 to at least the 1930s); König & Wernicke GmbH of Waltershausen (manufacturer, from 1911 to the present day, of all types of doll, including dolls' house dolls); Hermann Krauss of Rodach bei Coburg (manufacturer between 1925 and 1930 of papier mâché and composition dolls); Paul

Date 19th century
Maker Unknown English
Height 45in (114cm)
Width 36in (91cm)
Depth 27in (69cm)

The Tudor dolls' house was made in the early part of the 19th century as a replica of the Tudor manor house once owned by Henry Norris (or Norreys), who lived during Henry VIII's reign. The initials *H.N.* and the date 1509 may be seen on the right-hand gable. A portion of the roof may be raised at the back to give access to the interior, and the roof itself is oak, carved to resemble the solid slabs of stone used in Tudor times. Between the two gables, the front section may be removed to reveal the hall and upper landing, while access to the lower and first floor rooms is obtained by a separate doorway to each. The interior of this magnificent house, and the sad history of Henry Norris, were described in detail in an article in *The Connoisseur* in the 1920s by Mrs F. Nevill Jackson.

Courtesy: Bowes Museum, Barnard Castle, County Durham

Date **1908–30**
Maker **Thomas Batty**
Height **50in (127cm)**
Width **63in (160cm)**
Depth **29in (74cm)**

The Batty dolls' house is a unique model, begun in 1908 and planned and constructed by Thomas Batty over a period of 22 years. Mr Batty, of Drighlington, near Bradford, Yorkshire, originally started the model for his future grandchildren, but, sadly, his only son was killed during World War I, and he concentrated on making it as an exhibition piece, determined that it should not be a mere toy for children to play with, but a work of art and example of the work of a master craftsman.

The house is constructed in painted pine. It is of semi-classical design, having a double pediment slate roof with a central doorway with marble steps and a balcony above. The front may be removed in three sections to reveal a four-roomed interior together with a hall, landing and stairwell. Each room has windows to the front and sides, and is wired with two electric lighting systems.

The hall has an inlaid marble floor and a double staircase that leads to a spacious landing. The dining room on the left of the hall has a finely built ceiling. The oak

furniture is upholstered in red morocco leather; the miniature bookcase is stocked with books; and the sideboard bears a decanter and glasses, a telephone, framed photographs and ornaments.

The drawing room to the right of the hall is elaborately designed in the style of Louis XIV. The furniture is finished in 22-carat gold leaf, and the table is inlaid with mother of pearl. The grand piano has eight octaves, and each key was separately cut from billiard balls and an ebony stick. Delicate paintings panel the walls.

The two bedrooms contain unusual features. The bed in the room above the dining room almost presages fitted furniture: incorporated in the impressive bed-head are two bedside tables and , alongside, corner dressing table sections. The other bedroom contains a bed made to Thomas Batty's own design: the bed-head is a combination of wardrobe, medicine chest and safe. The net curtains are on wooden poles of late-Victorian or early Edwardian style. The floors throughout the house are of finely designed parquetry.

When the house was auctioned in 1984, a proportion of the proceeds was donated to the NSPCC.

Courtesy: Phillips, London

Krauss & Co. of Nuremberg (manufacturer of wooden toys); William Kreher of Olbernau (manufacturer of kitchen stoves, utensils and metal toys); Gebrüder Märklin of Göppingen (manufacturers of kitchens, stoves and bathroom fittings including showers and lavatories); A. Marsching of Nuremberg (manufacturer of paper and card products and games); Georg Möller of Sonneberg (manufacturer between 1911 and at least 1930, of dolls of various types); Wilhelm G. Muller of Sonneberg (manufacturer between 1900 and at least the 1930s of all types of doll, including those made of celluloid); Emaille Industrie Nederlandsche N.V. of Gravenhage (manufacturer of slates and so on); Dora Petzold of Berlin (manufacturer, between 1919 and at least the 1930s, of dolls, including art and boudoir dolls); Emil Pfeiffer & Successors, Vienna and Sonneberg (manufacturer of bisque dolls, papier mâché items, including shops, some wooden toys and buildings from 1894 to at least the 1930s); Alfred Rank (Successors) of Weinböhla (manufacturer of many dolls' house items, ornaments and accessories); F. Reiser & Co. of Berlin (manufacturer of model railways and tin toys); E. Reuter of Blumenau (manufacturer of wooden building bricks, toys and so on); F.A.D. Richter & Co. of Rudolstadt (manufacturer of stone building blocks); Hugo Roithner & Co. of Schweidnitz (manufacturer of wooden games and toys); Scharrer & Koch of Bayreuth (made glass ornamental items, beads and so on); Eduard Schmidt of Coburg (famous for the patented Sicora [*Sico*] dolls, "Wunderpuppen" – "they walk and they talk" – and also made boy, girl and baby dolls between 1904 and 1932, when it took over Loffler & Dill of Sonneberg and continued to operate under that name); Clemens Schmieders of Blumenhau (manufacturer of toy pianos and musical novelties); Fritz Stadelmann of Rothenberg (manufacturer of Christmas decorations and novelties); Tietz & Pinthus of Nuremberg (manufacturer of cardboard games and novelty items); Bruno Ulbricht of Nuremberg (manufacturer of dolls' house furniture; Thos. Volk of Furth (manufacturer of metal tops and a variety of toys and metal objects); Heinrich Wagner of

Left-hand house
Date Late 19th century
Maker Unknown English
Width 32in (81cm)
The wooden walls of this house have been painted to simulate brickwork and stone quoining, and the roof is covered with printed paper simulating slates. There are two storeys and two sets of chimneys. Squared bay windows stand on either side of the front door, and there is a small fence and front gate. The house opens at the back to reveal three rooms.

Second from left
Date Late 19th century
Maker Unknown English
Width 39in (99cm)
The façade, painted to simulate brickwork, represents a pair of semi-detached houses, each with bay windows and arched windows in the gables. The back opens to reveal four rooms. The structure, with its fenced garden, rests on a turned wooden stand.

Second from right
Date c.1820
Maker Unknown English
Width 35in (89cm)
This Georgian cupboard house is of painted wood, and the fanlight above the front door and the five windows are also painted. The front may be removed and shows two storeys of two rooms only. The drawing room has a side door, and both rooms have a cornice, dado, skirting board and fireplace.

Right-hand house
Date Late 18th/early 19th century
Maker English
Height 28½in (72cm)
The Willmot house has both an unusual façade and a unique provenance. The house is also illustrated on page 37 where its provenance is described. The furnishings are illustrated on page 65.

The items in the foreground of this illustration are unconnected with the houses; they were in the same auction and photographed at the same time.

Courtesy: Christie's, South Kensington; photograph: A.C. Cooper

Grunhainichen (manufacturer of dolls' furniture and children's furniture: it is interesting to note that D. H. Wagner & Sohn, also of Grunhainichen, made wooden dolls' houses and furniture, in complete, boxed sets but that this company used another British agent, named Bell, to represent it); Werner & Schumann of Berlin (manufacturer of kindergarten toys and children's games); and, finally, Zoo Toy Works of Munich (manufacturer of wooden toys).

But now let us look in detail at some of the dolls' houses manufactured between about 1840 and the outbreak of World War II.

When it appeared at auction, the Willmott house's extraordinary provenance was published. The story, dictated by "Mumsie" on 14 August 1950, is worth giving in full. Mrs E.M.A. Garrett, Great Grandmama, was born on 6 July 1817, and the dolls' house was made by an old man servant of her father, William Willmott, for the older members of his family before Mrs Garrett's birth. The servant had started to make the dolls' house but contracted cholera, which swept London in the early years of the 19th century. He was thought to be dead and placed in his coffin, which was carried to the place of burial with, as was the custom, his master following behind on foot. The coffin was rested on a stile in the neighbourhood of the Old Kent Road, which was then in open country, and tapping was heard coming from inside the coffin. His master insisted that the coffin be opened, but the pall-bearers, fearing contagion, refused. Determined that the coffin be opened, the master sent them away to fetch tools and opened the coffin himself while the others took themselves away to a safe distance. The servant was found to be alive, and, when he had recovered, he finished the dolls' house he had started and presented it to his master's children. The old man never completely regained his strength and was known as "Spider" to the end of his days.

The house, which is illustrated below, is as fascinating as its provenance. The wooden walls are painted to simulate brickwork with stone quoining. There are three bays and two storeys, with a circular oriel

FROM A PATENT, No. 18,225, DATED 9 SEPTEMBER 1899, TAKEN OUT BY SUSAN DIANA ROSA OF EDINBURGH FOR IMPROVEMENTS IN DOLLS' HOUSES

This invention relates to improvements in doll's houses and consists in forming the sides of the house so that they can be opened whilst the front of the house may be a fixture, so that a child can play with the dolls in the rooms and at the same time have the use of the street or front door for the dolls to pass in and out, and also have the front of the house with its windows intact.

In order that the sides when open may not be in the way, they may be made in sections hinged together so that they can be folded into a small space and be secured at the back of the house by a releasing band or chain. When closed, the various sections may be held rigid by a cross bar hinged or pivoted at one end and engaging with a spring or other catch at the other.

The various apartments may be arranged and fitted in any usual manner.

The empty interior of the Willmott house. Note that the kitchen grate has been inserted upside down.

Courtesy: Christie's, South Kensington; photograph: A.C. Cooper

ABOVE, RIGHT:
Date **c.1870–80**
Maker **Unknown English**
Height **53in (135cm)**
Width **49in (124cm)**
This impressive house, Coburg, which is in the private collection of Mrs Graham Greene, is flanked by a kitchen block, 20×20in (51×51cm), and a stable, also 20×20in (51×51cm).

OPPOSITE:
The interior of the main part of Coburg house consists of six rooms and a superb panelled dining hall, which occupies the whole of the ground floor and is dominated by a fine balustraded double staircase.

RIGHT:
The kitchen of Coburg house has a cook and housemaid busily at work. Note the fly paper, complete with dead flies, hanging from the ceiling and the apron on its hook on the wall.
Courtesy: Mrs Graham Greene, The Rotunda Museum of Dolls' Houses, Oxford; photograph: Frank Newbould

Date **1910**
Maker **English**
Height **14in (36cm)**
Gracie Cottage is a hand-built miniature house. Built to a scale of approximately 1:36, it depicts a town house in the Georgian style. There are two storeys, and there is a pediment with a slate roof behind. The front and sides are painted with brickwork and creepers, cow parsley and foxgloves. The front opens to reveal two rooms with their original furniture, vases, floor coverings and watercolours.

Gracie Cottage was made by the hall porter at the Greenwich Infirmary as a present for the matron's daughter, and the maker's inscription is on the front of the house. Gracie Cottage stands on a base and is covered by a dome.

Courtesy: Christie's, South Kensington

window in the pediment. The front door has a semi-circular fan light and pediment, while at the sides are protruding central bays and painted windows. The left-hand bay forms a deep kitchen dresser; the right hand bay is a hinged, glazed entrance to the dining room. The house opens at the front in two horizontal sections to reveal three rooms. The drawing room has a corner fireplace, gilded cornice and painted dado. On the ground floor are a kitchen, with an upside down cast-iron grate, separated by an interior door and dividing pilaster, from the dining room, which has a painted tinplate fireplace partially obscuring the painted wooden surround. Each room has rubbed wallpapers and friezes, and the house is furnished with many handmade pieces, including an inlaid, drop-leaf side-table with shell design, a folding games table, a painted tinplate sarcophagus wine cooler, several hand-coloured prints, including one in a metal frame, and a set of five Regency dining chairs. There are also many copper items, including a funnel, coffee pot, fish kettle and two coal scuttles, and a Dutch oven and a plate warmer, patchwork rugs, painted tinplate bellows, two warming pans and six early Victorian metal toy plates.

Coburg house, which is illustrated overleaf, is a prestigious mid-19th-century house in the private collection of Mrs Graham Greene, and she has kindly provided the following information about this treasure.

The house was named "Coburg" because, not unusually for mid-19th-

century dolls' houses, the word was chalked on its roof. It had been in a depository for many years before being auctioned at Christie's, and it has so far proved impossible to discover the name of the previous owner or the attribution of the armorial bearings on the façade.

When acquired, Coburg house was unfurnished except for two chandeliers and one picture, all dating from the 1870s. The double staircase and most of the window frames were in fragments, but the fruitwood panelling of the large ground floor hall was intact. Layers of modern paper had to be removed, and the only original paper is in the master bedroom on the first floor.

Coburg house has been furnished as if it were an English hunting box owned by one of Queen Victoria's more obscure relations, a Grand-Duke of Hohenlohe-Langeburg or of Schleswig-Holstein-Augustenberg, and this explains the number of mementoes and photographs of the Queen. The period is c.1870–90. Some of the furniture is 19th century, and some of the pieces are modern replicas. The stars on the chapel ceiling are positioned as at the time and date of purchase – 20th April 1980. The scarlet silk curtains are part of a vestment bought in the Sunday market at Brussels, from a convent now closed. The coronet on the sheets was specially embroidered. The real stained-glass window was copied from an old church in Munich and was created by Mrs June Astbury. The stables contain horses made by "Julip"; each horse has its name on a brass plate over its manger – Teck, Balmoral, Sefton and Albany.

An impressive yet at the same time sensible large mid-19th-century English dolls' house is Abbey Grange. This charming residence (see pages 42 and 43) is on display at all times, but without the handsome façade. The kitchen is on the left-hand side of the hall (most houses of this period have the kitchens in this position, and many had cavernous basement kitchens). This particular kitchen is equipped with every possible contrivance and has a cheerful, unhurried atmosphere, a wooden Grödnertal doll sitting by the bottle jack and hastener as if overcome by the heat. Across the hall, on the right, is the dining room. This room is furnished in a mixture of woods, which combine without offence because the eye is immediately attracted to the choice mahogany pedestal table, a lovely piece of exceptionally fine quality. The pedimented bookcase is well stocked with books. The pale, gold-embossed velvet fully upholstered settee and matching chairs look invitingly comfortable, suggesting that the dining room did double service as a small parlour. Abbey Grange is a commodious abode and has a butler's pantry, reached by the communicating door from the dining room. A dresser against the back wall is filled with "pewter" and silver plates, and there are polished pine tables on which are displayed a multitude of the "pewter" and silver items usually found in such a household.

The stairs lead to the middle floor. On the left is the small upstairs parlour (usually taken over by the mistress of the house as a sanctum) and through the connecting door is the master's study or smoking room. The square piano in the upstairs parlour bears the words, stencilled in

Date Late 19th/early 20th century
Maker Christian Hacker & Co.
Height 37in (94cm)
Width 27½in (70cm)
This mansard-roofed house in the French style was made by Christian Hacker & Co., a firm of toymakers in Nuremberg at the end of the 19th and beginning of the 20th century. The company, which was founded in 1870, used as a trademark the intertwined initials *CH* surmounted by a crown. This house was made for the French market. The front façade may be opened to reveal four rooms, a hall and a staircase, and the roof, which is covered with grey lithographed or painted paper, is also movable to allow access. Note the quality of the painting and the attention to detail of the decoration around the windows and attractive door. A feature of Hacker manufacture is blue line decoration on cream-painted furniture – on the kitchen dresser, for example – and fireplaces were fitted as standard in many of the company's houses. (See also p. 24)
Courtesy: Sotheby's, London

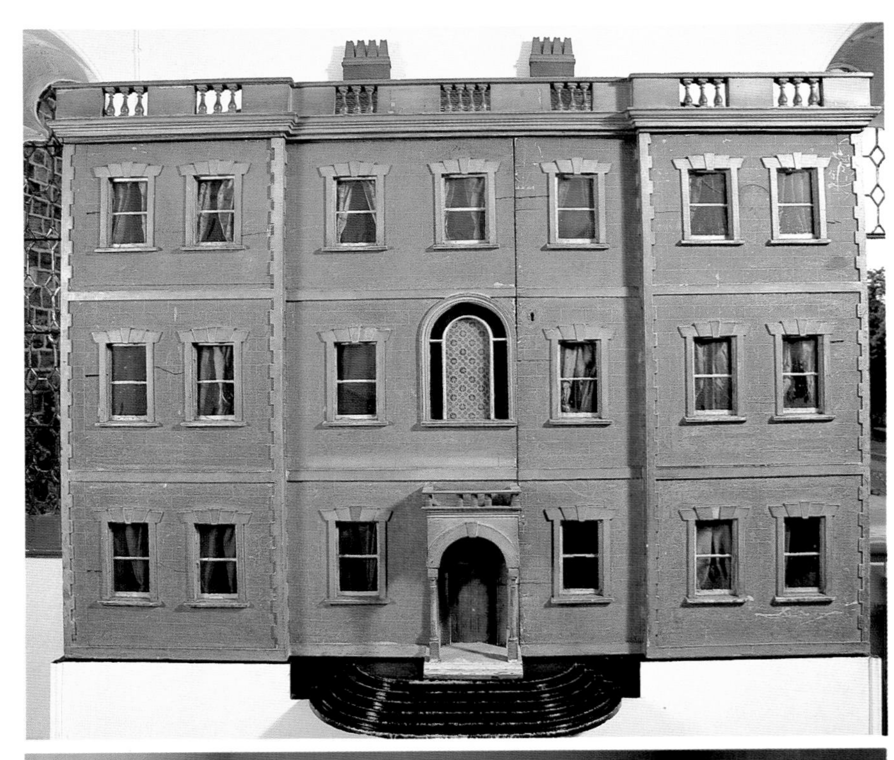

ABOVE, RIGHT:

Date **Mid-19th century**
Maker **Unknown English**
Height **48in (122cm)**
Width **65½in (166cm)**
Depth **19in (48cm)**

Abbey Grange is a splendid mid-19th-century dolls' house. It has three storeys and 10 rooms and a hall with a fine staircase leading to two landings. The mellow exterior has a "sandstone" finish, quoinings, balustrades and an impressive portico, above which is a tall arched window of stained and frosted glass. The other 19 windows are fully glazed. I should like to record my thanks to Paul Larkin and Alan Garlick of the Abbey House Museum, Kirkstall, who arranged for the façade to be brought from storage to be photographed for this book.

RIGHT:

The bathroom at Abbey Grange is 9¼in (23cm) wide. The bath, in addition to the two taps, has an overhead shower fitment; all of these function from a water tank behind the "tiled" surround.

Courtesy: Abbey House Museum, Kirkstall, Leeds; photographs: Frank Newbould

ABOVE, LEFT:
The Abbey Grange dolls' house seems to have absorbed some of the peace and tranquillity of its surroundings, and the rooms, with their internal doors and linking staircase, give an impression of a well-loved family home.

LEFT:
A close-up of the seldom-found stencilled inscription *Waltershausen b/Gotha* on the piano in Abbey Grange. When my photographer Frank Newbould and I were photographing the house, I was allowed to inspect boxes of items in the store room of the Abbey House Museum to look for suitable items to photograph. I found the square piano in a box labelled "miscellaneous Victorian dolls' house furniture" and remarked to Mr Alan Garlick, the Assistant Curator who was accompanying me, that, as it was Waltershausen, it would match the furniture already in Abbey Grange. It was not until we saw the piano in good light that the stencilled words became visible. Similar but unmarked square pianos do appear from time to time, but they are not plentiful and they are expensive. To see a marked example is a collector's dream!

Courtesy: Abbey House Museum, Kirkstall, Leeds; photographs: Frank Newbould

gilt, *Waltershausen b/Gotha* (the *b* an abbreviation of *bei*). This is extremely rare. Dolls' house square pianos do appear from time to time, but they are not plentiful, and unmarked, but otherwise identical pianos, are expensive. Across the middle floor landing is the oak-panelled drawing room, which is furnished *à la mode* with the Japanese black and gold lacquered furniture, the bright red contrasting strongly. (This type of décor became very fashionable in the mid- to late 19th century, but it was more brash than the delicate and restrained cabinets and chinoiserie furnishings of the 18th century.) Nevertheless, in a dolls' house this furniture is highly esteemed by collectors, but, as in all things, you either like it or not. I *do* like it and would most certainly buy such a complete suite if the opportunity presents itself. The magazine rack of soft metal contains a miniature copy of *Punch* magazine.

Another flight of stairs leads to the top floor. The door on the left opens into the nursery, the connecting door of which leads to mamma and papa's bedroom, where there is a German twin-bedded suite of golden pine and a choice side chair with wooden frame in simulated walnut and hand-embroidered upholstered back and seat. Across the landing, the other bedroom contains more Waltershausen furniture and a handsome half-tester bed and full-length dressing mirror. Again a connecting door leads to the bathroom, which is complete with shower, bath and pine washstand and various toilet articles. If you cannot easily get to the Abbey House Museum to see this wonderful dolls' house and its contents, study the illustrations as closely as possible and admire every detail.

The Wyreside Park dolls' house is a well-preserved early 19th-century dolls' house, which was equipped by Mary Jane and Louisa Garnett

Date **c.1884**
Maker **Benjamin Hazen Chamberlain**
Height **38in (96cm) (including cupola)**
Width **43in (109cm)**
Depth **20in (51cm)**
Illustrated here is the richly furnished parlour of the Chamberlain dolls' house. The house was made by Benjamin Hazen Chamberlain of Salem, Massachusetts, between April and December 1884, as a Christmas present for his two daughters, Mamie and Millie. Their names are inscribed on a silver nameplate on the front door, and Benjamin Chamberlain, a silversmith by trade, also made a little silver tea service for the dining room. He also made and decorated most of the furniture. The dolls in the house wear the original dresses, which reflect the styles of the 1880s. I learnt much of the history of this lavishly built house from the indefatigable former curator of the Wenham Museum, Mrs Elizabeth Donaghue, who had met Miss Emily Chamberlain, the Millie mentioned above, in 1955.
Courtesy: Wenham Historical Association and Museum, Wenham, Massachusetts; photograph: Robert E. Crosby

between 1835 and 1838. (The actual Wyreside Park House, near Lancaster, was burned down but was rebuilt in 1820, and the dolls' house remained at Wyreside Park until the house was sold in 1937.)

Two illustrations of the interior of this important dolls' house are shown, the first was kindly supplied by Christie's of South Kensington, the second, by Sotheby's of New Bond Street (see pages 64 and 69). The purpose of the two illustrations is to show an historical "before and after" example of a dolls' house interior as it *was*, fully furnished and complete with all the original furniture, numerous adornments and thirteen dolls

BELOW, LEFT:
Date **1986–8**
Maker **Celia Mayfield**
Height **40½in (103cm)**
Width **18½in (47cm)**
Depth **10½in (27cm)**
The façade of the Mayfield house, which is discussed in detail on page 51.
Courtesy: Celia Mayfield; photograph: Roland Cowen

ABOVE:
The interior of Mayfield house and, left, the dining room.
Courtesy: Celia Mayfield; photographs: Roland Cowen

This selection of Waltershausen furniture from the Abbey Grange house includes the rare square piano with a stencilled inscription in gold lettering: *Waltershausen b/Gotha* (the *b* stands for "bei"). The piano is 3in (8cm) high, 6¼in (16cm) wide and 3in (8cm) deep.

Courtesy: Abbey House Museum, Kirkstall, Leeds; photograph: Frank Newbould

when it appeared at Christie's – and later, when it was sold by Sotheby's as a separate lot with vacant possession, the furniture and furnishings split up into various lots, and even the dolls' house dolls were sold separately. One takes a delight in seeing a house sold with all original contents intact knowing (or hoping) that a piece of history in miniature will remain so, and will go on to delight collectors for many years to come. Alas, this was not to be for the Wyreside Park dolls' house, but I do hope that you will enjoy the record of the original interior. The contents will go on to entrance collectors who, having acquired a dolls' house of similar period but devoid of furnishings, furniture or dolls, leap at the opportunity to buy items for a well-loved, perhaps rather forlorn, dolls' house to give it a new lease of life.

Collecting dolls' houses and appurtenances spans continents and countries, and nations' products were introduced into several countries

Date **1986**
Maker **Kevin Mulvany**
Height **54in (137cm)**
Width **66in (168cm)**

Britannia House was designed, built and furnished and decorated by eight well-known British designers and auctioned, in February 1988, in aid of the African Medical and Research Foundation (AMREF).

The wooden house was based on the proportions of a Robert Adam building, with the façade details of stonework, plastering and brickwork finished to a high degree of accuracy. The interior and exterior lights are linked to a transformer concealed within the roof cavity and operate on both European and American voltages. The back of the house may be removed to reveal six furnished rooms, a hall and a landing, each decorated by a different English interior designer to a scale of 1:12. The dining room was decorated by Lavinia Dargie of Dargie Lewis Designs Ltd. It contains a circular table with eight Chippendale-style chairs, a decanting cradle and decanter, a pair of tea-caddy table lamps, a circular drinks table, a screen, an overmantel mirror, a fireplace with fittings, a tapestry over a serving table, and a jardinière, log basket, some miniature porcelain jars and a chandelier.

The library was decorated by Caroline Racher for Nina Campbell Ltd. It contains a two-seater sofa, a butler's table, three upholstered wing armchairs, a desk, library shelves laden with books, a fireplace and accessories, a circular table with a lamp and photograph frames, a petit-point carpet, a chest of drawers and accessories.

The hall and landing, decorated by Mary Fox Linton and Peter Hunter from Fox Linton Associates Ltd, contains miniature garden ornaments. There are also a plexiglass table, a lantern and a mirror.

The music room, decorated by Bill Bennette of Bill Bennette Design Ltd, has a parquet floor and delicately painted walls. It contains a marble topped gilt console table, a fireplace with accessories, a harpsichord and a cello, a music stand, gilt circular tables, two salon chairs and four gilt and upholstered armchairs.

The drawing room, which was decorated by Fiona Merrit for Colefax and Fowler, has a parquet floor with a petit-point carpet. The display shelves are laden with miniature porcelain, and there are a two-seater sofa, upholstered in miniature chintz, with an armchair to match, another upholstered armchair, a padded footstool, a fireplace with an overmantel mirror and accessories, a card table with simulated Georgian chairs, a lacquered cabinet on a stand, a sofa table and other accessories.

Decorated by George Cooper of Cooper Perkins Ltd, the yellow bedroom contains a wing armchair upholstered in petit-point, a writing table with a chair, a firescreen with an embroidered panel within, a console table, a four-poster bed, bedside tables, and miniature accessories, including luggage.

The pink bedroom, decorated by Jo Robinson from Mrs Monro Ltd, has floor boards partially covered by a petit-point carpet. It contains a bed, painted side tables, an Empire-style rattan day-bed, a writing table with a simulated Georgian chair, an upholstered armchair with matching foot rest, a chest of drawers, miniature porcelain dishes hanging on the walls, lamps and other accessories.

The nursery was decorated by Diana Hanbury of The Tarrystone. The walls were painted with a jungle mural by Rupert Gatfield, and the floor boards are covered with a petit-point carpet. There are a sofa, chairs and tables, and the shelves are laden with toys and other amusements.

The carpets throughout the house were made by Patricia Borwick and Cynthia Jacobs.

Courtesy: Sotheby's, London; photograph: Traditional Interiors

at the same time. During the 19th century, German manufacturers turned out large quantities of dolls' houses. Many of these houses had the same basic structure but were modified and adapted by the manufacturer. Christian Hacker of Nuremberg, for example, is known to have introduced alterations to architraves, roofs, doors and so on, and to have made other slight variations to appeal to the importers of his products and to enable them to amend their catalogues from time to time. European manufacturers appointed agents in different countries, including America, and that often meant the main stores in America did not have direct dealings with the makers of the goods. This has led to a paucity of information about the source of these products, even though pictorial evidence of German, French and English products appears in the catalogues described as "the latest imports from Europe".

In American collections I have seen some unusual dolls' houses, which are named "mystery houses", and the makers are unknown. Most museums own one, and there are several variations in the Margaret Woodberry Strong Museum. They vary in size, but all have a common feature – ornamentation consisting of wooden strips applied to corners of the exterior of the houses. My own view is that this decoration is a combination of quoining and rustication. They are like no other houses, and there is no architectural term I have been able to trace that accurately describes this feature. I suppose one could almost call them "hybrids". The same ornamentation occurs around the base and sometimes under the windows and elsewhere.

Flora Gill Jacobs' authoritative book *Dolls' Houses in America* gives facts and theories and includes excellent illustrations of some of these "mystery" houses.

As in most countries, not all miniatures in America are imports. Many

Date **c.1900**
Maker **Unknown Swiss**
Height **50in (127cm)**
Width **44in (112cm)**
Depth **19½in (50cm)**

This extremely rare and large, open-fronted dolls' house was made by a craftsman in western Switzerland at the turn of the century. At the left are the words "Villa Poupée" and, below, the word "Magasin". Across the front of the gallery is the legend: "L'ordre et la propreté, la patience et la gaieté sont vertues cardinales pour mes filles les principales. [Order and cleanliness, patience and cheerfulness are the cardinal virtues for my daughters.]"

This splendid house has an open front, galleried across the upper floors, and the tiled roof is partly hinged for access. The windows are of various designs, some containing stained glass. There are four rooms and a kitchen in the main part of the structure: the lower floor has a kitchen and dining room, while on the upper floor are the drawing-cum-living room, the nursery and a bathroom. In the extension to the left are a bakery on the ground floor, a schoolroom on the intermediate level and an intricately constructed staircase providing access to both the veranda above and the main building.

Each room has wallpaper and carpet of a different pattern, and the rooms are lavishly equipped and furnished. Many highly collectable items emphasize the character of each room from the kitchen to the bathroom, from the parlour to the nursery. There are 31 *mignonnettes* (dolls' house dolls), some all bisque, some part bisque and some celluloid.

Courtesy: Christine Kohler, Auktionhaus Ineichen, Zürich; photograph: Margie Landolt

are the results of creative people, craftsmen and commercial manufacturers, in whom America is rich. There are craftsmen and -women turning out exquisite miniature furniture and equipment, room settings and so forth.

When one thinks of American dolls' houses, one's first thoughts are of Bliss, Converse, McLoughlin and Schoenhut, but studying the wonderful and unique creations reveals much about the background and combination of cultures – cooking equipment and stoves used in various regions of the country, different forms of heating, porches and stoops. The more one studies the interior of these houses, the more evident it is that one is seeing American domestic life as lived in the 19th century.

Finally, a brief mention of modern dolls' houses. Craftsmen and companies still produce fine examples, of course, but recently I have been impressed by what can be achieved from the kits that are now available.

Aided by a carpenter, Celia Mayfield successfully adapted a kit to suit her own design. The furniture, dolls, furnishings and numerous other artefacts in the Mayfield house (see illustrations on page 45), many of which were created by Mrs Mayfield herself, were brought together in just two years, between 1986 and 1988. One year after Mayfield house was completed Celia had a brainwave and sub-divided some of the original rooms. The study, for example, now has its own *en suite* bathroom. Later, Mrs Mayfield added the top storey with three extra rooms. One of these was turned into a well-equipped housekeeper's room, with linen in the cupboards and an ironing table. Great imagination and invention have been shown throughout the house – an old birdcage was dismantled and parts of it used to make a playpen for the nursery, fireguards, fenders, jardinieres and other items; pictures from tea packets adorn the walls; magazines have been made from advertisements for books.

Mrs Mayfield's passion is for building and creating all things miniature from the dolls' houses themselves to intricate pieces of furniture, from miniature food for tiny kitchen tables to miniature plants for balconies and parlours. Many of her houses have recessed fireplaces behind which is concealed a small red electric light to give a welcoming glow to the room. So successful has Mrs Mayfield become that she now makes dolls' houses, furniture and so on to order. Some of the items are combinations of kits, some are her own ideas.

When one sees the results that have been achieved by miniaturists such as Celia Mayfield it becomes easy to understand why this hobby is enjoying such popularity and gaining new enthusiasts so rapidly. Their creations, whether of period or of modern houses, allow us to escape from the rigours of modern life and step into a different world.

3
DOLLS' ROOMS AND SHOPS

Manufactured in Germany c.1890, this painted wooden room setting with turned pillars comprises a drawing room and a bedroom. Both rooms have floral wallpaper and printed paper simulated tiled floors. In the drawing room are a set of black and gold painted furniture in the art nouveau style, a piano, a jardinière with potted plants, various ornaments and, in the foreground, a white metal (tinplate) stove. The water tap is Austrian and was made in 1900. Note, too, the three-fold screen of metal and card, which bears a label reading "Carl Quehl, Nurnberg". The bedroom furniture is painted white. The room is 35in (89cm) wide.

Courtesy: Christie's, South Kensington; photograph: A.C. Cooper

Model rooms are known to have been made by both craftsmen and commercial organizations, albeit in small quantities, from the 17th century in Germany, Holland and France, and, in the form of Presepios, in Italy. Kitchens were the most popular. Styles varied slightly from region to region, the plenishments of the early Dutch and German kitchens being the most elaborate – the Dutch kitchen at the Museum of Childhood, Bethnal Green, London, and the Nuremberg kitchen at Bowes Museum, Barnard Castle, Co. Durham, are two fine examples. Contemporary engravings provide evidence that children did indeed play with kitchens so that young girls could assimilate a knowledge of kitchen equipment and culinary utensils before they were married.

From the early 18th century onwards, shops were popular toys. Grocery stores, daintily furnished milliners' shops complete with fancy bonnets and other fripperies, butchers' shops, made in both Britain and Germany, dairies and farms were all widely produced and played with. But more than that, for these items also played a part in the school curricula for young children in the 19th and even the early 20th century.

In the 19th and 20th centuries, room settings, shops of all types, and

kitchen equipment increased in popularity, although interest waned during the Depression. Nevertheless, cheaper versions of these toys continued to appear right up to World War II.

In this chapter we shall look at dolls' rooms and shops and, finally, at the miniature silver items that were made to decorate and embellish houses and rooms.

DOLLS' ROOMS

Some of the earliest baby houses were intended as educational models rather than frivolous toys. Anna Korferlin, wife of a burgher of Nuremberg, who furnished a large cabinet house in about 1631 and was perhaps the first person in history to charge the public a fee to view a dolls' house, made the instructional intent of the house clear when she wrote about it in a pamphlet for visitors. "It will give you a good lesson, so that when you return to your home, or when God gives you your own home, you will be able to order your lives and organize the duties of your household rightly."

It was because their main purpose was to teach domestic skills through play that the earliest rooms modelled seem always to have been kitchens. Although these earliest kitchen rooms are still known as "Nuremberg kitchens", it is at least possible that they originated in the Netherlands rather than in Germany.

In 17th-century France miniature sets of rooms became popular, but not all of them seem to have been instructional. Early in the century the daughters of Henry IV owned several model rooms, including a lying-in room – lying-in rooms seemed to have had a particular appeal. In 1630 Cardinal Richelieu, Louis XIII's first minister, presented one to the young

Date c.1700
Maker Unknown German
Height 21½in (55cm)
Width 30in (76cm)
Depth 16½in (42cm)
German kitchens were the finest ever produced. They provided both instruction and play for children – the pewter plates and plenishments being scaled down versions of human-size kitchen equipment. Children did play with them and ate tiny portions of food and biscuits from the miniature plates.

This German toy kitchen was made c.1700 in Nuremberg. There is a similar fine kitchen on display in the Bethnal Green Museum in London.

Courtesy: The Bowes Museum, Barnard Castle, County Durham

Karl Träger Patent, 15 April 1903.

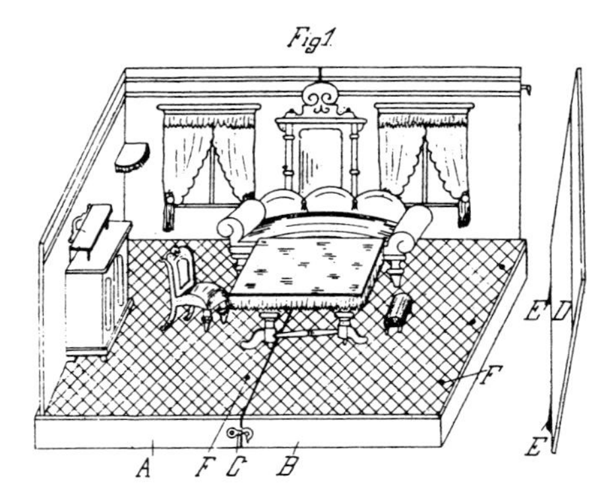

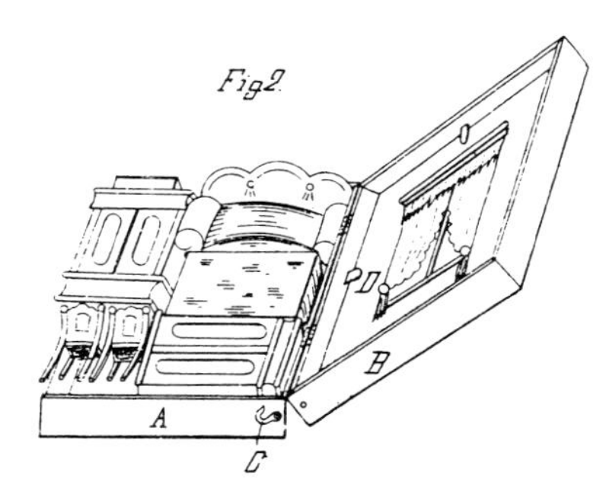

Room settings of this type were shown in store and commercial catalogues throughout Europe and in Britain and the United States. Examples may be seen in several museums (charming rooms in the French style, with luxuriously draped windows and best quality furniture, may be found in a number of museums in the United States), and they even occasionally come up for auction.

Collectors, especially beginners, are always looking out for fresh ideas, and a room setting would not be too difficult for an amateur to make. If it were made to suit the age and period of some already owned pieces, it could become a suitable home for a set of furniture, some accessories and perhaps two dolls.

FROM A PATENT No. 115,858, DATED 3 JULY 1917, TAKEN OUT BY TOY MANUFACTURER DAN LIONEL COLTEN OF LONDON FOR IMPROVEMENTS IN TOYS

This invention relates to improvements in toys and has for its object to provide a toy constructed of flat units which are put together by a simple form of detachable connecting means such as has been proposed for articles of furniture, and the invention particularly relates to the means for securing a hinged unit such as the front of a doll's house to one of the adjacent flat units and to the specific construction of the whole article.

The form of joint adopted is similar to that proposed for sectional furniture in which a screw turn-pin is secured to one part, which turn-pin passes through a slot in the other part and is then turned so that its head lies across the slot and prevents the two parts from being separated.

According to the present invention, the hinged unit in a toy, the units of which are connected as stated above, is attached by hinges to a batten, which batten is engaged with the adjacent flat unit by turn-pins in the same manner as the units are held together.

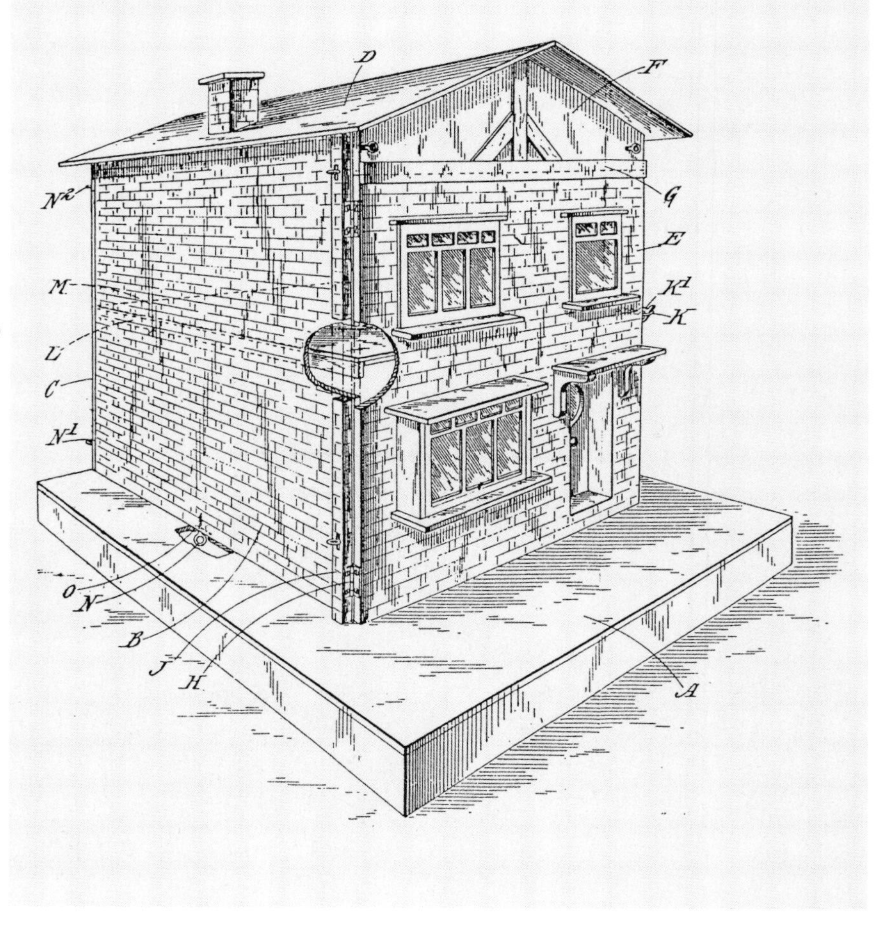

Princess d'Enghien, who was permitted, so the anecdote goes, to play with its mother, grandmother, midwife, nurse and maidservant dolls, but not to bath the baby, which, being made of wax, would have been irreparably damaged by the bath water. This room was almost certainly a political gift in return for favours received – the princess was a member of the powerful Condé family, whose support Richelieu needed to retain.

The French followed the lead set by the "Nuremberg kitchens" and made imitations in the German and Dutch style before beginning to make kitchens in their own individual regional styles. Model rooms remained popular in France until the 20th century, although, as the years passed, they dropped down the social scale. The Musée de l'Histoire de l'Education in Paris has a sumptuously furnished bedroom of about 1850, more complete than many dolls' rooms in that it has both a top and a glass, opening, front.

FROM A PATENT, No. 164,251, DATED 22 JUNE 1920, TAKEN OUT BY THOMAS SPENCER, AN ARCHITECT, OF MORTLAKE, LONDON, FOR A NEW OR IMPROVED DOLLS' HOUSE

My invention consists in an improved dolls' house of the type composed of a number of separate components which can be readily assembled into one complete model without the aid of any tools.

According to my invention the components which form the walls, floors, etc. are rigidly secured in position by means consisting of simple tongue and slot clips permanently attached to or integral with them. The completed model may be extended by the addition of similar interchangeable parts to form a block of two or more doll's houses etc. When not in use the model can be dismantled and the components packed flat in a shallow box.

A folding toy dress shop. The three sides of the interior of the sales room are lithographed in bright colours and show great attention to detail. Four life-like paper dolls are included, together with the table bearing four bonnets. When not in use, the whole may be folded back into the original box with its pictorial label, which bears the name of the toy in three languages – "New Magazine of Mode/Neues Mode Magazin/Nouveau Magasin de Mode" – which suggests that the toy is of German origin and dates from the 1840s or 1850s.
Courtesy: Phillips, London

Miniature room settings crossed the Atlantic to the United States of America at least as early as the beginning of the 19th century. Clark, Austin and Smith were producing dolls' rooms by the middle of the 19th century and they were followed by Bliss and McLoughlin. The line between folding rooms and folding dolls' houses here, though, becomes indistinct. More than one American manufacturer sold sets of four rooms, formed by assembling four slotted pieces of card to make the internal walls, raising the question, when does a room set become a house?

The firm of Schoenhut began to manufacture wooden "toy apartment house rooms" soon after 1917. Matching furniture could be bought separately. Production of these rooms continued until 1934.

These, of course, were toys, but the 20th century has also seen the return of the 17th-century European idea of using rooms as display cabinets for collections of miniatures or as dioramas showing period furniture and decorative styles – like museum rooms in miniature. In the Thorne Rooms at the Art Institute of Chicago is one such modern collection. There are here 68 model rooms, all to 1:12 scale.

The rooms encompass almost every conceivable style and period from the 16th century to c.1940. Skilled craftsmen and -women worked to ideas formulated by Mrs James Ward Thorne on her many travels abroad. Some are meticulous miniature reproductions of special rooms in palaces, country houses, chateaux and salons while others show typical rooms representing many American periods and styles, from a Shaker living room to the parlour of a tavern in Connecticut.

In Britain, the fine Carlisle Collection was formed by Mrs F.M. Carlisle, a dedicated collector, whose collection of some 28 miniature rooms of miniature furniture and artefacts and genuine antiques is rivalled only by the specially commissioned furniture and room settings, which were

The outer case containing the Chippendale Library is approximately 40in (101cm) wide, 13in (33cm) high and 21in (53cm) deep. The Library itself, made in England in the 20th century, forms part of the Carlisle collection of miniature rooms now housed at Nunnington Hall, near Helmsley, North Yorkshire. There are more than fifty volumes in the breakfront bookcase, including an Almanack of 1770 and copies of *The Koran* and the *Bhagavad Gita*; also included is the message broadcast by Queen Elizabeth II on her Coronation Day. Among the books on the drumtop table are a leatherbound album of photographs of the Carlisle family and a visitors' book containing signatures (in miniature) of all the contemporary craftsmen whose work contributed to the excellence of the remarkable collection of which the library is part. The carpets are executed from original Persian designs, and the maps on the walls are by Pieter Keer and are dated 1599.

Courtesy: National Trust and Nunnington Hall; photograph: Ken Shelton

made by some of Britain's finest craftsmen. The Chippendale Library (see page 56) must suffice to whet your appetite for the entire collection, which is housed in Nunnington Hall, near Helmsley, North Yorkshire, and is now in the care of the National Trust. Mrs Carlisle created and stitched most of the carpets and upholstery in petit point. She commissioned many craftsmen, who worked to a scale of 1:8, notably F.J. Early, who is well known for his creations for Queen Mary's dolls' house. J.W. Thomas made the magnificent musical instruments. The collection also includes a wonderful Palladian hall built, in part, to emulate the hall of Hatch Court in Somerset; the miniature version of the hall and the furniture were made by Albert Reeves of Peckham Rye. Mrs Carlisle personally embroidered all 88in (224cm) of stair carpet. There is also a superb Carlton house desk by F.J. Early in the Regency games room. The collection is, quite literally, stunning, and a corner cabinet at Nunnington Hall houses some wonderful genuine pieces of antique miniatura – furniture, porcelain and so forth.

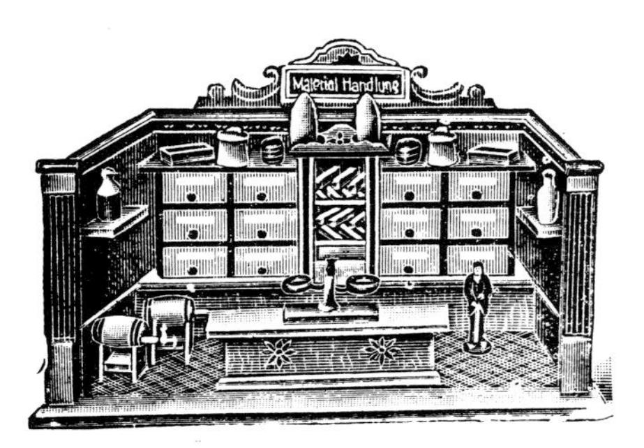

DOLLS' SHOPS

Separate miniature shops have existed in their own right probably as long as miniature rooms. In 1696 the 14-year-old Dauphin of France, Louis, Duke of Burgundy, was reported to have owned "nine shops of the market place, filled with little pieces of enamel". The Dauphin's shop may really have been a market stall. Model stalls have a very long history, separate from that of shops, and they may have preceded shops proper.

Bestelmeier's illustrated catalogues, of between 1794 and 1807, describe a number of toy shops. In France, some years further on into the 19th and 20th centuries, *épiceries parisiennes* were sold in a variety of sizes. Practically every major country produced toy shops stocked with a wide range of regional produce. By fairly early in the 19th century toy butcher's shops were being made in England. Why they should have become popular is not at all clear. Certainly many shops were playthings with an additional educational purpose.

Some of the most elaborate early 19th-century shops were made in Germany. The Toy Museum in Nuremberg has three early particularly delightful examples. The earliest – made like a dolls' house, with an opening front – is a milliner's shop. It dates from about 1820. The hats are displayed on papier-mâché mannequin busts, and other fashion accessories – reticules, purses, and handkerchiefs – are on sale. Behind the counter is the wooden-doll milliner, wearing a bonnet. From around 1858 is a shop selling basketware – there are miniature shopping baskets (many of them with decorative weaving and some with painted decorations), laundry baskets, and a cradle on wheels. The third of these fascinating shops was made at the very turn of the century. It differs from the others in that it has no back and no sides; it is simply a wooden counter with a wooden storage cupboard behind it. It is a general store, selling everything from sweets to starch, from peppermints to petroleum.

In the last half of the 19th century and even well into the 20th century,

PREVIOUS PAGE:
A well-made butcher's shop with living accommodation above. The carved wooden joints of meat and the carcases are realistically painted in red and white. Shops such as these were made by many manufacturers in the Nuremberg area during the second half of the 19th century. This example is 24in (61cm) high, 20in (51cm) wide and 13in (33cm) deep, and it is not marked. There is another particularly good example in the Bethnal Green Museum, London. The duck in the foreground, hurrying to safety with her brood, was made by Lehmann; the horse-drawn bus, which is of unknown manufacture, is marked *Enterprise General des Omnibus*.

Courtesy: Abbey House Museum, Kirkstall, Leeds; photograph: Frank Newbould

FROM A PATENT, No. 15,670, DATED 6 NOVEMBER 1915, TAKEN OUT BY GEORGE PRYCE WALLINGTON OF CHELTENHAM FOR A CONSTRUCTIONAL TOY

This invention relates to a constructional toy whereby from the parts provided and herein described all kinds of model buildings such as half-timbered houses of one or two storeys or cottages, halls, hospitals, etc. true to scale and rigid in structure, can be erected upon a given base and taken to pieces.

According to this invention I provide a base board, which may be made to fold into two or three pieces and may form part of the box or boxes in which the various parts are stored.

In this case I bore holes about ⅜ inch diameter at predetermined distances along the lines of the plans of the various model buildings to be erected thereon, these holes are provided at each return angle along these lines and also at other intermediate distances suitable for the doors, windows and walls of the buildings; the distances apart of these holes are made very exact to allow of the interchangeability of the various parts.

I provide uprights of ⅝ inch or other suitable square section, of various lengths, tanged at each end to fit into the holes in the base; these are provided with grooves running longitudinally on 2, 3 or 4 sides as may be required.

I provide panels of thin wood or cardboard or other suitable material,

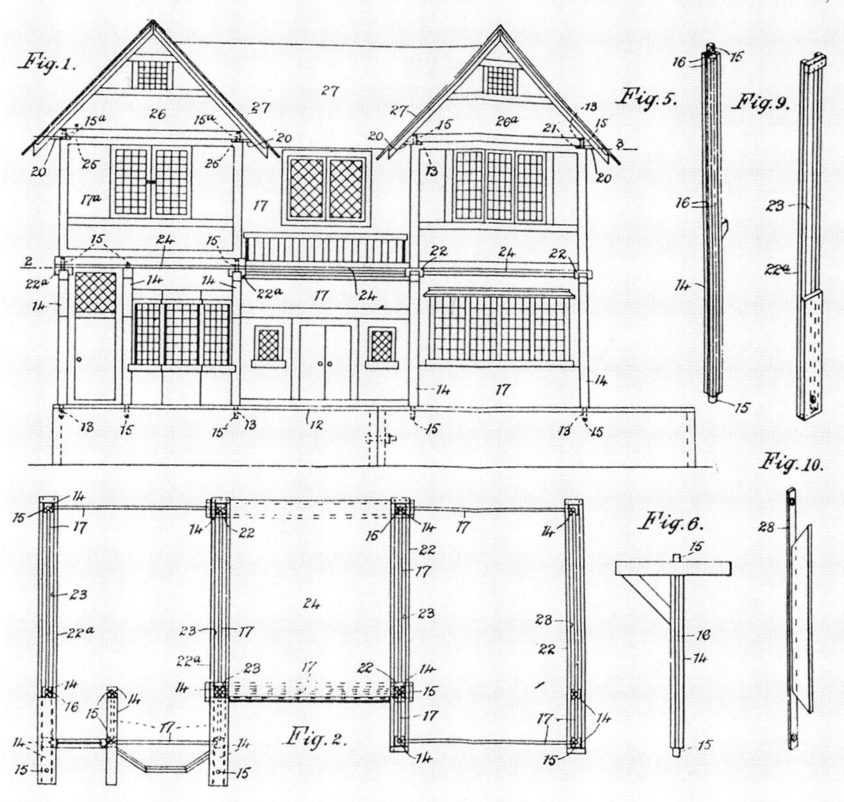

This extremely rare grocery and "Colonialwaren" was made in Germany c.1870. The detail lavished on the fittings, drawers, labels, tins, bottles and produce is a joy to behold.

Courtesy: Auktionhaus Ineichen, Zurich; photograph: Margie Landolt

ANIS
NELKEN
VANILLE
ZIMMT
MANDELN
ZITRONAT
ZUCKER
NUSSE
Essig.
Pflaumen.
Birnen.
Zucker
Gemüsesuppe
Bismarck
Maggi Würze
MAGGI's
Fleischbrüh-
Würfel
Riebeck
Kerzen
MANDELLEBKUCHEN
TH. HÄUTLE
MÜNCHEN

of such widths that they will slide in the grooves of the uprights, when these are erected in the holes in the base, specified for the particular building which is being erected. These panels are painted on one side to represent the outside wall of the house and on the other side to represent the interior wall of the room.

Or in place of such panel, I provide an open frame to slide into the same grooves, this frame is capable of holding a loose panel which may be removed to enable the inside of the room of the house to be viewed.

I also provide doors and windows of various designs, tongued on each side, to slide into the smaller openings provided by the intermediate holes before mentioned.

To tie the structure together, I provide tie plates made of strips of wood or card board or a combination of these substances, having holes in them corresponding with the holes along the lines of the base; these tie plates fitted on to the tangs at the top of the uprights, after the panels have been inserted, tie them together and form a rigid structure.

In the case of one or two storey buildings I further tie them at each floor by means of miniature channel girders provided with square holes in the web which fit over the uprights, allowing the girder to rest on the top of the panel of the floor below; this ties one set of uprights together and these are tied to the next set of uprights by means of the floors which rest on these girders and engage into them.

folding shops made of lithographed card were available. The best of these were finely detailed, brightly coloured dioramas representing three sides of the shop's interior; they were sold with accompanying paper-cut-out customers. Ordinary lithographed-paper-on-wood German shops continued to be made until World War I and in the United States of America the Bliss range of toys also included shops of various kinds. In Britain Raphael Tuck produced examples suitable for all ages and pockets.

MINIATURE SILVER

Historical records from the 16th, 17th and 18th centuries reveal that elaborate and expensive rooms were created for members of royal families, and high quality silver and gold and delicate porcelain items were made in miniature by many of the leading craftsmen of the time. In 1576, for example, the daughter of Henry II of Germany commissioned silver "buffet pots, bowls, plates and other silver toys such as they make in Paris" as a gift for the children of the Duchess of Bavaria. The mother of Henry IV of France had a plate inventory in which is listed "a doll's set of silver plenishments set with diamonds". English nobles were not to be outdone. In 1952 Christie's of London sold a miniature goblet, attributed to c.1630, and silver miniature items bearing London hallmarks of the 17th century prove that not all such miniature silver items were imported.

Baby houses were commissioned to grace large country houses and to provide interest and genteel occupations for the ladies of the household, who were expected to pass their time embroidering miniature furnishings and so on. For the kitchens and drawing rooms of such houses, however, miniature silver items were needed.

FROM "PARIS REVISITED" BY GEORGE SALA

Going down to the great toy shops of the rue Vivienne, the rue St Honoré, and especially the *Enfants Sages* in the Passage Jouffroy. I found the Easter Egg losing its luxurious, losing its decorative, but retaining a recreative, and asserting a practical character. What do you think of an egg containing a complete *batterie de cuisine*, pots and pans *fourneau économique* and all? An egg holding a complete *mobilier* for a doll, chairs, tables, sofas, cabinets, looking glasses, bed, bedding, likewise attracted much attention in the *Enfants Sage*, as did also an egg which served as a receptacle for a complete parlour photographic apparatus, an egg full of gymnastic appliances; and an egg on which being opened, disclosed – a baby doll in her cradle.

Surviving examples of the miniaturist's art are true reflections of the styles of their particular periods. Furniture, tankards, tea equipage, chocolate pots, punch bowls, mirrors, miniature candlesticks complete with snuffers, silver replicas in miniature of animals and pets – all these and more were made to entrance and delight. The first London silversmith to specialize in silver miniatures may have been George Middleton, a descendant of Sir Hugh Middleton, the celebrated goldsmith and jeweller to James I and Charles I, but many other silversmiths produced silver toys and replicas of full-size items. Miniature silver was sold in London in Deards' shop, and Lady Mary Wortley wrote in her *Farewell to Bath* in 1736: "Farewell to Deards' and all her toys which glitter in her shop." John Deard died in 1731, but many miniature silver items were in the Westbrooke collection and bore his mark, *ID*.

The earliest miniature silver was finely constructed and executed, most being formed from flat plate. The finest details were not overlooked and as much care was taken as for full-size items. In the 19th century much of the miniature silver was cast, but some of the more finely wrought items were still made from flat plate. Other notable silversmiths were Augustine Courtauld, Edward Dobson, Anthony Ellines, Joseph Lowe, Isaac Malyn, John Sotro, Viet and Mitchell, and Wetherell and Janaway. The Westbrooke baby house possessed miniature silver items bearing the mark *CL* (the trademark of John Clifton, Foster Lane) and hallmarked for 1718. Two chairs in the late 17th-century style bear the mark of Matthew Madden of Lombard Street, which was registered in 1696.

Sharp-eyed collectors of miniature silver should keep a look-out for the marks on any miniature items in their collections – darkened silver items have been spotted by keen-eyed sleuths in the most unlikely places.

4
DOLLS' HOUSE FURNITURE

Dolls' house furniture was, and still is, made all over the world, and the many styles produced commercially, as well as those made in limited quantities by craftsmen, including those working at home, would fill a book ten times the size of this one. This book is primarily concerned with the period from the beginning of the 18th century through to the outbreak of World War II. An enormous variety of materials was used to create furniture and appurtenances for dolls' houses during this period, including bone, horn and ivory; card- and fibreboard; ceramics; enamelled materials; feathers, fabrics and even fish bones; gilt ("Tiffany" style); metals ranging from painted and lacquered hard metals, including tin (as made by Rock & Graner, for instance), to the softer filigree type (known as "Diessen"); paper; pins, cord and cork (making miniature furniture from these materials was taught in kindergartens); tortoiseshell; and silver and pewter-type "silver". There was the high-quality, Waltershausen furniture

The Wyreside Park house opens in two sections at the front to reveal a central staircase and hall, a landing and four rooms, each with a fireplace, one with a gilt overmantel. The kitchen has a cooking range, and there are fitted corner shelves on the landing. The wallpaper and flooring are contemporary throughout.
Courtesy: Sotheby's, London

The original late 18th- and early 19th-century furnishings of the Willmott house, including a games table bearing the label "W. Vale's Tunbridge Ware & Toy Rooms, 62 Fleet Street". In the near foreground is the rare painted sarcophagus wine cooler. For a full description of this fine house and its contents, see pages 36–7.
Courtesy: Christie's, South Kensington

RIGHT, ABOVE:
Date **c.1835–8**
Maker **Unknown English**
Height **72in (183cm)**
Width **65in (165cm); 75in (191cm) with stand**
Depth **23in (58cm)**
The façade of the Wyreside Park house shows the two-storey house painted cream and beige to simulate stone, with two chimneys on the "slate" roof. The house has a chained balustrade at the front and sides. At the front are five painted windows, and the central front door is flanked by pillars under an arched portico. There are two real windows on the left-hand side and two painted windows on the right.
Courtesy: Christie's; photograph: A.C. Cooper

RIGHT, CENTRE:
Three miniature metal items from the Wyreside Park house. On the left is a three-wheeled pram; on the right is a high-backed chair with a blue velvet seat; and in the centre is a high chair, with a lace-covered pink cushion. As an indication of scale, the high chair is 3½in (9cm) high.
Courtesy: Sotheby's, London

RIGHT, BELOW:
Further items from the Wyreside Park house. In this lot (only part of which is shown here) were five unframed prints and one framed print, two framed photographs, one unframed photograph and one framed watercolour. The two framed photographs, one of Queen Victoria and one of the Prince of Wales, have figured gilt frames, 1¾×1½in (4.5×4cm), and two wooden frames (one of which may be seen at the left) are decorated with leather leaves, are 2¾×3¾in (7×9.5cm) and contain country and seaside scenes with people in a landscape and a church. On the right is a three-panelled screen with prints of children contained in soft metal filigree frames; each panel is 5in (14cm) high and 2½in (6cm) wide. There were also two painted silk face screens, a cream checked silk fire screen on a turned metal stand (the screen is 4in (10cm) high), a turned metal hat stand, 5in (14cm) high, with four hats including a pink silk bonnet, and a hanging gilt metal flowerpot.
Courtesy: Sotheby's, London

of simulated rosewood, gilt-stencilled and ormolu-applied wood (which was given the name "Duncan Phyfe" by Vivien Greene in her book *English Dolls' Houses*), and other finishes included chinoiserie, Japanning and lacquering. Sometimes wood, cherrywood or pine, was left plain or painted shades of cream or in white. Oak furniture was manufactured by the short-lived company Elgin of Enfield (see page 126). Eric Elgin, together with his two sisters, formed a company in 1919 to produce this better-than-average commercially produced English dolls' house furniture. He provided employment for many ex-servicemen and subsequently sold much of his output to Lines Brothers. This furniture, without the impressed Elgin mark, was marketed by Lines under its trademark *Triangtois*. It was sold in sets as separate items under the generic name "Jacobean", and it was seen in Lines catalogues as early as 1921. Elgin's company closed down in 1926, and after the factory's closure, Lines made furniture to some of Eric Elgin's original designs.

We must not forget, either, the gaily coloured, lithographed wooden furniture that was made in Germany and exported all over Europe and to the United States. In America, Bliss made the A.B.C. line of lithographed furniture, and Schoenhut made wooden furniture until the company closed in 1935. In Britain, Raphael Tuck, perhaps better known for greetings cards, made furniture and also dolls' houses and even dolls from card.

Plastic furniture is not discussed in this book, but good-quality furniture is still being made today, by the vast army of craftsmen and -women working at home and also by companies, which offer commercially-made furniture that ranges in quality from items for the finest reproduction "mansion" to the more modest dwelling. The choice is yours, but, as far as possible, refrain from mixing "antique" or "made yesterday" items yourself. If your finances or circumstances dictate that this must be so, don't lose sleep: remember that in most real houses today

ELGIN
ENFIELD

The trademark of Elgin of Enfield. Look for this mark under well-made English furniture dating from the 1920s. The furniture is made with great attention to detail in a darkish "brown oak" finish. (see also page 81)

This collection of metal and tinplate dolls' house items, also from the Wyreside Park house, dates mostly from the mid- to late-19th century. The long-handled warming pan is copper, and there were two copper and one black-painted tin coal scuttles. Other fireside paraphernalia included three pairs of tongs and two shovels, a fire guard and a gilt-metal, pierced fire surround. In the foreground is a soft metal sweetmeat basket. At the back of this group is a cream, orange and gold painted wall-mounted water cistern, 5½in (14.5cm) high.
Courtesy: Sotheby's, London

SAMLO
INSTRUCTIONS
FOR OUTFITS A TO E
PRICE 6D
SAMLO
FURNITURE
DINING ROOM SUITE

PRECEDING PAGES:

RIGHT:

The interior of Wyreside Park house complete with the original furniture, dolls and fittings.

Courtesy: Christie's; photograph: A.C. Cooper

ABOVE LEFT:

This most desirable turned and painted wooden dinner service is decorated with blue lines and flower sprays. The original box shows that the manufacturer was A. Lerch and that the set was made in Germany, probably in the late 19th century. It has proved impossible to discover further information about A. Lerch.

Courtesy: Christie's, South Kensington; photograph: A.C. Cooper

BELOW, LEFT:

Samlo constructional building sets to make houses, garages and other buildings in a variety of styles were patented by H.A.F. Sieck, of 104 Hulgaardsvej, Frederiksberg, Denmark. The patent, number 443,055, was dated 6 February 1935, and the registered trademark number was 564311. The patent read: "A toy building set . . . comprises cardboard units having trapezoidal and rectangular tongues and recesses along their side edges for inter-engagement . . . the roofs may be fitted with foldable chimney units and step units. . . . The buildings may incorporate gables and dormers and may be of pentagonal or polygonal form. . . . Window frames of different colour from the other units may be fastened in openings in the wall units . . . the buildings may have a flat roof formed from the floor units."

The building sets, instruction booklets and other items of paper and cardboard were produced to the above patented specifications by John Waddington Ltd, Wakefield, Leeds. The Danish wooden furniture was assembled from coloured and polished wooden sections slotted into grooves. The kit shown in this illustration is a dining room suite, complete with paper simulated parquet floor on top of which rests a paper simulated Persian carpet. Kits for bedroom suites and for kitchen suites were also made.

Author's collection; photograph: Frank Newbould

the old and the new often live harmoniously together. It is all very well being a purist, but prices today are so high that many collectors will have to choose between leaving rooms, or even whole houses, empty or contriving to put reproduction pieces in the darkest corner. Some other subterfuge may be considered – turning a plainish house into an "antique-cum-second-hand-shop", with all the oddments of furniture on display, and even priced, as if in a shop.

A dolls' house empty of furniture may appear deserted. The point is well illustrated if we take the case of the Wyreside Park dolls' house, which is illustrated on pages 64 and 69. Novice collectors will learn much from this example if they ever find themselves in a similar position and have to re-equip a dolls' house. Ideally the furniture, furnishings and inhabitants should be of the same era as the house.

Christie's catalogue described the house as "furnished and equipped with country made and manufactured pieces". These "pieces" included an English upright piano; a Gothic-style, mirror-backed sideboard; kitchen and dining-room tables and chairs; a writing table; a stained and inlaid chest of drawers; a double bed with bedding and shaped ends; a dressing mirror; two corner cupboards; free-standing shelves; a plate rack; a circular table; a painted tinplate, three-wheel perambulator; a painted tinplate wall cistern; a turned wood dressing-table mirror; a stained bone games table, candlestick, and inkstand; painted turned sets of plates; copper kitchen equipment; a soft metal screen with coloured prints of children; a shoe cleaning brush; a hot-water can with bucket and bath; and a pair of hand-painted fringed silk fans. There were also 10 china-headed dolls' house dolls. Four of these were "elaborately dressed" with gilt paper decorations, lace and ribbons, one as a man in Turkish costume. One doll with a jointed wooden body and china limbs was dressed as a man in a black suit, and three Grödnertal dolls had moulded and painted elaborate hairstyle, jointed wooden limbs and contemporary clothes.

All this and more in five rooms.

The history of Wyreside Park house tells, in reverse, what most dolls' house collectors try to do: to fill their house with appropriate and contemporary furniture, fittings and dolls.

This example shows that, for the collector, dolls' house furniture is both a delight and a difficulty. The delight, of course, lies in the patina and quality of the early 19th-century country-made items – chairs, tables, sofas, cabinets, beds and miscellaneous household pieces – in perfect miniature. The difficulties are manifold. Dolls' house furniture exists in almost infinite variety (as do the houses themselves), so that it is not always easy to follow the historical progression. Some of the furniture is craftsman made, some commercially made, and some home made. Some of it is not even dolls' house furniture at all, but odd items pressed into service. At one end of this scale are the superb apprentice pieces, at the other, items made by handymen or children.

The tradition of dolls' house and miniature furniture and furnishings

See also pages 36, 71, 122 and 123

FROM AN APPLICATION FOR A PATENT, No. 148,871, DATED 10 JULY 1920, MADE BY THE FIRM OF BRUNO ULBRICHT OF NUREMBERG FOR A METHOD FOR THE MANUFACTURE OF DOLLS' FURNITURE

This invention has for its object, to create pieces of toy- and doll's furniture which are almost true imitations of the original pieces of furniture and sufficiently strong not to break easily. The invention has been stimulated by the well known bedsteads of iron construction with wooden panels in the head and the foot part. The bedsteads made of angular- or otherwise profiled iron or of tubular rods are of a light appearance although they are of great stability and resistance but this construction has hitherto not been applied to the manufacture of other pieces of bedroom furniture such as wardrobe, washing-table and the like, as such furniture would not look well at the side of other wooden furniture of more elegant and softer outlines. It seemed impossible to use this metal construction for the manufacture of doll's furniture by simply reducing the measurements. According to this invention it has become possible to apply the metal construction to the manufacture of toy- and doll's furniture by using strips of sheet metal for making frames for the front and back parts of the furniture including the feet and top ornaments. . . .

The invention is not limited to the manufacture of bedroom toy-furniture but can also be applied to the manufacture of furniture for drawing rooms, kitchens and the like.

In the accompanying drawings a doll's bedroom furniture is shown by way of example, which has been made according to the invention.

flourished in all countries. Children were encouraged to be creative in the schoolroom and at home. The peak of activity was reached, probably, in England in Victorian times. Mrs Beeton, for example, gave the advice reproduced on pages 74–5 for those attempting to furnish a doll's house in the home.

In the 17th century it was not a toy maker but a silversmith who made the miniature silver playthings so loved by the aristocracy. One by-product of this was the extensive use of precious and semi-precious metals in early baby houses. In this and the next century some of the most exquisitely beautiful pieces of dolls' house furniture of all time were made in silver. Almost everything needed to equip a miniature house was crafted out of silver – chairs, tables, kitchen implements, whole fireplaces – and complete house furnishings were made as well as single pieces, which were sold separately. Gold furniture was also made, albeit less frequently. Pewter was used extensively, especially for kitchenware; copper and brass were used for utensils in many Nuremberg kitchens and copper was used also, for other dolls' house plenishments.

The great period for commercially made dolls' house furniture was the 19th century and up to the outbreak of World War II. Germany, France, Britain and the United States all made furniture and exported it. Because of the complexities of the export trade between the countries, dolls' houses tended to be furnished in a variety of different national styles, far more eclectically than real houses of the period.

OPPOSITE:
Date **c.1880**
Maker **Unknown English**
Height **36in (91cm)**
Width **24in (61cm)**
Depth **18in (46cm)**
Norwich house, an English cupboard house made of oak, has one glazed door and three rooms. When I inherited this house, it contained superb furniture from a house of an earlier period (1830–40). Having donated the furniture to a more worthy dolls' house, I decided to re-equip it with some of the furniture and fittings I already possessed to show that a three-roomed cupboard house can provide a home for items of differing periods and styles.

The Thonet bentwood furniture in the lower room and the other furnishings had been displayed in various makeshift room settings, as had the silver furniture, French cabinet and various *objets d'art* in the central room. The silver fireplace and matching chairs is a "one-off" set that had long awaited a suitable home. The delightful cabinet-made suite of Edwardian bedroom furniture was a lucky find. It was made in 1908 and the tallest piece is 8in (20cm) high. It appeared at an auction that advertised "and dolls", and I had been in two minds about going to another auction some distance away that had advertised dolls' house furniture. At the last minute, I decided to go to the nearer auction, the one with "and dolls", and found the added lot. The suite is scrupulously fashioned out of walnut, which has been lightly wax polished but not stained. The wardrobe has a fitted interior of hanging space and three side drawers and one bottom drawer; the washstand has a simulated marble top and two drawers and two cupboards below; the chest has two short and two long drawers; the dressing table has a swivel mirror and two short and two long drawers; and there is a double bed. All have lemon-coloured silk behind the intricate fretwork. A chair, a blanket box and a bed-side table complete the suite. The auctioneer later told me that it had been made in 1908 especially for the lady who had brought it in for sale. The ornate brass ceiling decoration in the bedroom is original to the house, as is the black cast iron and gold fireplace in the lower room.
Author's collection; photograph: Frank Newbould

ABOVE:
Examples of home-made and commercially manufactured furniture dating from the later Victorian period to the 1910s. The four-poster bed and the pin/cord/card table and chairs are home made. The shell-encrusted items are believed to be of English manufacture. The desirable red plush upholstered suite is German; note the leather inserts in the arms of the settee and armchair, which, in real life, would be rubbed away by hands and elbows.
Courtesy: Abbey House Museum, Kirkstall, Leeds; photograph: Frank Newbould

The German manufacturers sometimes copied furniture designs from other countries. For example, Paul Leonhardt of Eppendorf openly advertised dolls' furniture made to English "Jacobean" designs that were practically identical to Elgin furniture. Always look for the Elgin mark before paying a high price for a single piece. Filigree furniture from Diessen, and glass and pottery items from Thuringia were their own creations. Germany also produced perhaps the most famous of all dolls' house furniture – that now described as Waltershausen furniture. Most was made of imitation rosewood or ebony with imitation gilt, upholstered in silk or covered in red, imitation-leather paper, but it was made in such numbers and such variety that it is difficult to pin down. Early Waltershausen furniture is sometimes called "Beidermeier", after the real-life style of furniture on which many of its designs were based. Vivien Greene is responsible for the term "Duncan Phyfe", which is now in common use and applied to Waltershausen furniture – again because of the design style they reproduce in miniature. Waltershausen furniture was imported into Britain between about 1830 and the early 20th century.

The French began to manufacture dolls' house furniture on any scale

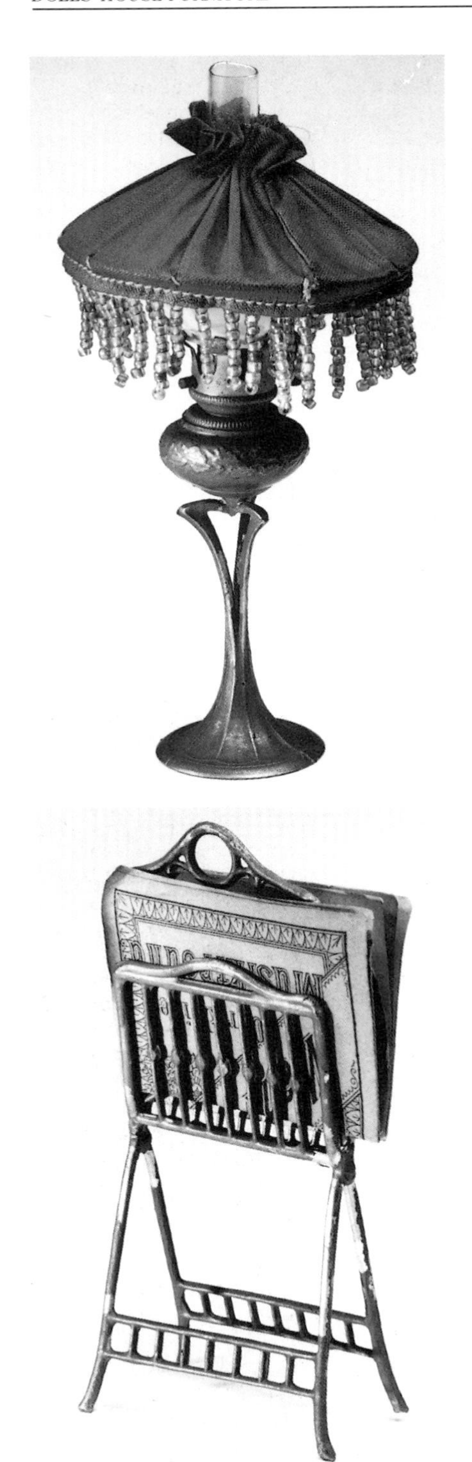

ANOTHER TYPE OF HOME MADE FURNITURE

Another simple way of making doll's furniture is as follows: Get some black perforated cardboard and some thin polished cane. For the window-curtains . . . a tiny pole of black cane studded with gilt knobs is required. If the little pole is not ready made, it can be quickly done by pushing the ends of a bit of cane into two gilt beads. The curtains are of white net darned in stripes with white *glacé* thread, and having vandyked edges worked in buttonhole stitch. For the lambrequin, cut two pieces of cardboard . . . and sew the two pieces together with long stitches of red filoselle. At the corner of each vandyke are small tassels made of red filoselle. The frame through which the pole passes is of four thicknesses. It is joined to the lambrequin on the side nearest the window. The curtain bands are of twisted or plaited filoselle with tassels of the same. The buckle is a strip of black cardboard stitched with red silk. Two small gilt nails fix the bands to the wall. The blind is of fine white muslin, neatly hemmed at the sides and edged with very narrow lace. The only way to draw this up is by making a wide hem to the top, and putting a thin piece of cane through the hem. This must be set on the top of two nails knocked into each end of the window. It will be very easy to roll or unroll the blind when wished. . . .

For the sofa the cardboard is of four thicknesses, and of oblong shape. It is then sewn with red filoselle in long stitches, both the length and breadth of the cardboard, forming a checked pattern. Now pierce holes through the four corners with a stiletto, and take a long piece of cane, which must be pushed into the two holes at the back to form the frame. Two more holes are pierced in the back, and through these another piece of cane is pushed and curved. It is fixed to the first piece by a small white stud in the middle. The second piece of cane furnishes the legs. For the front legs and arms two more pieces of cane are pushed through the holes in the front of the seats, and curved to meet the back, to which they are fixed with tiny nails. The cushion is also of cardboard, worked with red filoselle, with tassels of the same. For the easy chair we must make a stand of cane like an American chair. The two front pieces are shorter than the back. The seat and back of the chair are of cardboard, bent to . . . shape . . . with square vandyked edges. They are in one piece, and ornamented, like the seat of the sofa, with red silk. At the corners of the vandykes are small silk tassels. Holes are pierced through the cardboard at the top of the back and where it is bent for the seat part, and the longer pieces of cane are put through them. The shorter pieces are then fixed to the front of the chair in the same manner. These two canes are then crossed, and a third piece goes straight between the legs of the chair, and all are secured with white studs. The making of the table is very easy: fourfold cardboard . . . and ornamented with long silk stitches. The cane legs are put in in the manner already described, and secured with a stud. For the chair the cardboard is ornamented with the checked pattern of stitches already described. For the back of the chair one piece of cane is used, forming the legs and the curved back. The two front legs are pushed through holes in the cardboard, and headed with a white stud. The footstool is made in the same way. The carpet is of fine red cloth with pinked edge. It is embroidered over canvas in cross-stitch with black silk. When the embroidery is finished the threads are drawn out.

The whatnot . . . consists of three cardboard shelves, each four thicknesses, stitched with red silk, and having holes pierced at each corner, through which the canes forming the poles are passed. For the wall-bracket . . . take a double piece of cardboard and bend it in the middle. Then cut the top . . . and work it with red filoselle. The shelf consists of a square piece of cardboard rounded in front and put in corner-wise. It is joined to the back with black silk. The lambrequin is cut in vandykes, and ornamented at each point with red silk tassels.

The wardrobe consists of six pieces joined together with black silk. . . . The whole is worked with red filoselle, representing panelling. For the hooks, pins can be stuck into the sides and back on the inside.

For the mirror frame . . . double cardboard with the glass glued on to it forms the back. The front is cut out to let the glass show through. This is ornamented with red filoselle in the Grecian key pattern, and joined to the other cardboard which has the glass glued to it with stitches the whole way round. The console attached to the mirror consists of four thicknesses of cardboard. In the middle of the top is a small square hole by which the mirror is hung to the wall.

This little table is made of an oblong piece of cardboard of four thicknesses, and with holes pierced at the corners for the fixing of the legs. It has a lambrequin of cardboard, worked in square vandykes with red filoselle.

Doll's Work-Case. This is a small cardboard box divided into three compartments, and having a glass lid. It is fitted up with a pair of tiny scissors, a needlecase, and a thimble, mats for the table, &c., and wool of several shades with which to work them. Round the case are several little articles made of cardboard, a card-tray, a folio, waste-paper basket, satchel, curtain-band, trinket-case, date-case, and a needle-book.

only at the beginning of the 19th century. At first the pieces were copies of German styles. By the 1840s, however, France was producing its own styles of furniture in gilt metal, *faux bambou* or white wood with silk upholstery, resulting in a thriving export market.

Extracts from reports of exhibitions often give clues to probable makers. Wittich, Kemmel & Co. of Geislingen may have obtained orders at the Great Exhibition of 1851; they may, indeed, have been the manufacturers of the bone and ivory furniture illustrated on page 87. While German manufacturers were able to note the designs of British makers of dolls' house items, there was a reciprocal exchange of ideas at the exhibitions held throughout Europe. For this reason, it is difficult to make firm attributions of the countries of origin of many dolls' house furnishings.

Dolls' house furniture is always difficult to attribute. There is rarely a maker's mark on an individual piece of furniture; nevertheless, the boxes in which the furniture was packed were often labelled to show either the name of the manufacturer or the registered mark (see pages 116–31). By the middle of the 19th century tea and dinner sets for dolls were being made in vast numbers in Britain, Germany and France. Some were made of high-quality porcelain, others were of common or garden china and there is little hope of dating them except by external evidence.

Turning from commercially-made dolls' house furniture, we will now

OPPOSITE:
This rare art nouveau pressed-metal oil lamp has a glass bowl and funnel and a bead-fringed red silk shade. The pressed-metal music stand contains German sheet music. Both items date from the late 19th century.
Courtesy: Christie's, South Kensington; photograph: A.C. Cooper

OVERLEAF:
Date **1875–1900**
Maker **Unknown English**
Height **35in (89cm)**
Width **26in (66cm)**
Depth **14in (36cm)**
Issott house is a beguiling abode that does not fit into any category. It is undeniably a cupboard house, yet it has an unusual mixture of styles and periods. One theory is that the French salon (the central room in this three-storey house) and the hall and staircase were built a decade or so before the other rooms, including the bathroom, which is reached by a spiral staircase. It is thought that the later rooms were built by a different hand, which is borne out by their contrast in style to the elegant salon.

The master's study, on the left of the hall, which is not illustrated, houses a collection of bronzes and other artifacts. The French salon contains many items reminiscent of the past – two wax jacks, gilt cabinets, a portrait on an easel and some French light wood furniture, upholstered in blue silk. A Waltershausen secretaire is the only piece of German furniture in the room, although the chandelier is German.

The larger of the two bedrooms, which are not illustrated, has a small half-tester bed in the corner, and a Waltershausen davenport, dressing table and dressing glass. The upright piano is of German manufacture. The smaller bedroom is the children's nursery and contains Diessen metal cradles and two types of high chair and a French sofa. The dolls have Parian, china and bisque heads, but all have calico-stuffed bodies, except for the master, who has a composition body.
Author's collection: photograph: Frank Newbould

FURNITURE BY G. THONET, FROM THE *ILLUSTRATED LONDON NEWS*, NOVEMBER 1851

Our engraving represents a number of articles by G. Thonet, of Vienna, and gives a fair idea of Austrian furniture. Its great peculiarity consists in its simplicity and strength, together with its being adapted for packing, through the backs of the chairs and the front legs being removable. The durability of this furniture is mainly owing to the wood of which it is made being subjected to a process by which it is rendered elastic, and thus almost indestructible, while its rigidity is yet maintained. At a period of its manufacture the wood is capable of being bent into almost any shape – of being coiled into volutes or bent into ogees; but when thus bent it is fixed in the desired position, yet it retains permanent elasticity to such an extent as to render it incapable of serious injury by a fall.

The backs of the chairs and the hind legs are formed of one piece of wood bent into the form of an elongated horseshoe, and of some simple filling, and by this formation a great advantage is gained, as the grain of the wood runs throughout the length of the legs and back, and in no case has a transverse or cross direction. We have seen these chairs thrown for yards in the exhibition without sustaining injury, and jumped on in a manner calculated to test their strength severely, and yet they have been unhurt.

In design these chairs vary, the back being in some cases open, in others filled with scrolls or volutes, and in others with canework; the armchairs, the sofa, one small chair, and the rocking-chair in our engraving are illustrations of the latter form. The rocking-chair is exceedingly comfortable, and is a thorough luxury; and some of the other chairs are little inferior to it in this respect. No furniture could be better calculated for rough work than this, it being especially adapted for the nursery and playroom.

Date **1850–65**
Maker **Unknown English**
Height **54in (137cm)**

This painted wooden dolls' house with a galleried flat roof has been painted to simulate stonework. A fine house, it is unusual in that it has 16 windows at the front and 15 windows at the back. The front opens in three sections to reveal six rooms, a central staircase and landings, and interior panelled doors. The front door has a brass knocker and handle. The original furniture includes kitchen items, a metal swing cradle, a Waltershausen fitted dressing table, three bisque-head dolls, one of which is in its original maid's uniform, a Waltershausen stencilled piano and a five-panelled metal screen covered with scraps of pictures of children.

Courtesy: Christie's, South Kensington

look briefly at "oversize", cabinet and apprentice pieces. The furniture in the Norwich cupboard house (see page 73) and the apprentice pieces in the foreground of Two Gables (see page 80), might seem oversized for many dolls' houses; the latter would, in fact, be perfect for a house custom made to fit to a wall or a wardrobe house – i.e., a substantial old wardrobe or cupboard with suitable adaptations to make a large cupboard house or a variety of rooms. There are two interesting examples of this. The Voegler house in Chester County Historical Society, West Chester, Pennsylvania, dates from c.1836. It is 7ft 10in (2.4m) high, 6ft 1in (1.9m) wide and 2ft 8in (81cm) deep. The furniture is all apprentice-piece size. The superb large cabinet house was made by a Philadelphia master cabinet maker, Voegler. The dining room walls are wainscoted in part and, above, are painted landscapes, which are reminiscent of the Dutch cabinet houses, although there the resemblance ends. The façade of the Voegler house has three pilasters, a pediment and opens by means of two glass doors. There are three contemporary dolls, the largest of which is 10in (25cm) high, which raises the question: "when does a doll cease to be classified as a dolls' house doll?" If one takes the Voegler house as an

The bentwood Thonet furniture exhibited at the Great Exhibition in 1851 caused a great furore. Not only was it a novelty, it could be roughly treated and not come to any harm, an attribute that was actually demonstrated to the reporters. Bentwood furniture was subsequently made both for children and for dolls' houses, and dolls' house furniture bearing the label "Thonet Wien" has been found. The house illustrated on page 73 contains a set – table, settee and four side chairs – of Thonet furniture. Japanese imitations of the Thonet dolls' house furniture came on the market at the end of the 19th and during the early years of the 20th centuries.

Four "foxy red" painted cast metal and tinplate dolls' house chairs with shaped back stretchers, painted black "plats" (seats) and turned front legs. These chairs were made c.1840, and furniture in this style was made in both England and Germany at this time. These were made by the English manufacturer, Evans & Cartwright.

Courtesy: Christie's, South Kensington; photograph: A.C. Cooper

Date **c.1902**
Maker **Unknown English**
Height **30in (76cm) (to chimneys)**
Width **44in (112cm)**
Depth **17in (43cm) (between gables: 26in (66cm))**

Two Gables has an individually tiled roof, and the lower part of the house still has its original rustication. The upper part has been repainted in black and white in mock-Tudor-cum-stockbroker style. This heavy wooden construction was made by a carpenter; it is not, as might be thought at first glance, one of the commercially made Lines houses. Six sections may be removed from the front façade; another six move from the back. The dark oak back door opens directly into the kitchen, and another door leads from the kitchen into the hall. There are eight rooms in all. The hall is panelled in dark wood and has a delft rack and a plate rack on one wall. The staircase turns at the top and enters a small passage before opening out to a wide landing, which also has panelled walls and a delft rack.

The miniature sample kitchen range to be seen in the foreground is of the type to be found in real houses of this period; it is 7in (18cm) high, 8in (20cm) wide and 5in (14cm) deep. The grate is complete with heavy fire bricks and the oven door opens to reveal shelves. The mahogany chest is inlaid with a lighter colour; it is 6¾in (7.5cm) high, 8½in (21cm) wide and 3¾in (9.5cm) deep. The oak refectory table is 12in (31cm) long and 4¾in (12cm) high and wide; it is laid with simulated pewter tableware and there are two matching benches. Each of these three pieces – the chest, the table and the cabinet – bore, when bought at auction, a card reading: "This very important apprentice model was made by an apprentice whilst serving his time with the well known firm of Warings of Lancaster, later known as Waring & Gillows. This piece is authentic." The Waltershausen dressing glass is shown to demonstrate that, good though it is, in no way can it compete with apprentice pieces of this quality. The parlour of Two Gables, opposite below, is mostly furnished with Lines suites of "dark oak" with simulated leather upholstery. Originally this was black too, but I repainted it with an "aged" red, which I feel looks more realistic. There are three pieces of marked Elgin in furniture in the upper room.

Author's collection; photographs: Frank Newbould

RIGHT, ABOVE:
This Waltershausen simulated rosewood furniture dates from the 1860s and 1870s. It has the usual ebonized wood and ivory inserts and the gold stencilled adornments. The chest of drawers on the left has a rare miniature skeleton clock on it; second from right is a lady's *bonheur du jour*; at the right is a marble-topped dressing table with one long drawer and two cupboards. The games table with the folding top is a most desirable item; it is shown open, and the legs are extendable. On top of the table are a soda syphon and tiny Doulton jug. In the foreground is an unusual item: a painstakingly hand-made English chair in beadwork, a fashionable Victorian hobby, with a delightful embroidered cushion in the back of the chair.
Courtesy: Christie's, South Kensington; photograph: A.C. Cooper

RIGHT, BELOW:
This Waltershausen furniture is now in the bedroom of the Issott house. Dating from the 1860s, it is simulated rosewood embellished with gold stencilling and imitation ebony. There are ivory supports and turned wood decorations. The dressing table (left) has one long drawer, beneath which is a decorative wooden apron; it is 4¼in (10.5cm) high, 3¼in (8.5cm) wide and 1¾in (4.5cm) deep. The dressing glass, which is 4¼in (10.5cm) high, 3in (8cm) wide and 1¾in (4.5cm) deep also has a single long drawer. The davenport, possibly one of the most ingenious of miniature productions to come from Waltershausen, is 3½in (9cm) high, 3in (8cm) wide, 2in (5cm) deep at the widest point and 1¾in (4.5cm) deep at the narrowest point. There is a cupboard at the left-hand side and three drawers on the right; the lift-up desk lid has a simulated red leather insert. The ebonized chair at the right of the illustration is of unknown English manufacture. It is made in the Georgian style and has a faded red silk seat, edged with gold trim; it is 3¼in (8.5cm) high, 2¼in (5.5cm) wide and 2in (5cm) deep. The metal chandelier in the foreground is German and dates from 1870–80.
Author's collection; photograph: A.C. Cooper

example, one would say 10in (25cm), but many experts cavil at this figure and suggest perhaps an inch or two less. My own view is that an 8in (20cm) doll in one setting could look too tall and yet seem too small in a very large baby house. Each case must be judged on the overall effect – *you* must assess the proportions, and you can do this only if you take time to study and to proceed with caution.

The other example of a house or rooms that will accommodate antique apprentice- or cabinet-size pieces of furniture is situated in the Mary Merritt Doll Museum, Douglassville, Pennsylvania. Unfortunately, no illustration was available, but I was given permission to quote from the booklet, *Touring Mary Merritt's Doll Museum*.

> One whole wall of the museum is given over to small cases behind glass which show interiors with miniature fittings and figures assembled with the most accurate regard for scale. Chairs and tables fit the size of the doll, accessories are never over or undersized.
>
> The late Mary Merritt collected over a period of forty years, searching out fine quality and rare items, and the variety and time span of her collection are remarkable. At one end of the big museum room is the

RIGHT, TOP:

These four pieces of painted tinplate and metal furniture with "wood" finish were made by Rock & Graner of Biberach in the 1870s. The *bonheur du jour* with the filigree gallery stands next to a side cupboard, also with a galleried top; the inside of the cupboard is painted bright yellow, and the cupboard is 4¼in (10.5cm) high. The *étagère*, also with a filigree gallery around the top, has a filigree back panel behind a central mirror; it is 6in (15cm) high. The *secrétaire à abattant* also has a yellow-painted interior, and the cupboard below is disguised as three false drawers; raised on four bun feet, it is 6¼in (16cm) high.
Courtesy: Sotheby's, London

RIGHT, CENTRE:

The dolls' four-poster bed in this group is 8½in (21cm) high, and it was made in England in the last quarter of the 19th century. The sofa and four chairs, which are Waltershausen in origin, have chromolithographic decoration on the back and seats. They are 5¼in (13cm) high and date from 1850–70. The table is rare; made by Rock & Graner in 1855–75, it is 10in (25cm) long and made of tin, painted to resemble a grained wood.
Courtesy: Christie's, South Kensington

RIGHT, BELOW:

This miniature furniture was collected by its original owner between 1918 and 1939.

On the back row, from the left, are: a French *semaine* (seven-drawered chest) approximately 4in (10cm) high; a pedestal dining table with an extra leaf, 8in (20cm) long; a tallboy, the top cupboard enclosed by a pair of panelled doors, with two deep and one shallow drawers to the base, raised on cabriole legs, 6½in (17cm) high; a stained wood longcase clock, the removable hood surmounted by finials, 7in (18cm) high; a George III style stained wood breakfront bookcase with four glazed doors above, the base fitted with one long and two short drawers with cupboards below and fitted with a collection of simulated books, 8in (20cm) high; a stained wood corner cupboard with pediment, a glazed door above and cupboard in the base, 7in (18cm) high; a bow-fronted sideboard, a cellarette and cupboard to either side and a central drawer, raised on six tapering legs, 5in (14cm) wide; a day bed with a shaped headboard, 6in (15cm) long; and a bureau whose sloping, carved front encloses a fitted interior, with three long graduated drawers below, raised on bracket feet and approximately 3½in (9cm) high.

In the middle row, from the left, are: a bureau with a fitted interior and one short and one long drawer below, raised on cabriole legs and 3in (8cm) high; a Hepplewhite-style open elbow chair with pierced vase splat; a lowboy fitted with a dressing slide and one long and three small drawers above a shaped apron, raised on cabriole legs and approximately 2½in (6cm) high; a card table with a shaped outline and raised on cabriole legs, approximately 2¾in (7cm) wide; and a writing table, each side fitted with two short and one long drawer, raised on cabriole legs and some 5in (14cm) wide.

In the foreground are a pair of Hepplewhite-style standard chairs with pierced back splats; an upholstered wing armchair in the early Georgian style; a two-seater open Hepplewhite-style settee; and a miniature stool in the Queen Anne style raised on four feet. *Courtesy: Christie's, South Kensington; photograph: A.C. Cooper*

"Towne house", a group of four rooms – two bedrooms, a kitchen and drawing room with centre hall and stairs. Here the scale is larger. Pieces are all rare one-of-a-kind examples of 18th- and 19th-century furniture. With the exactly right accessories one can see vividly portrayed how the well-dressed doll from Queen Anne to Greiner lived.

There is a genuine 18th-century Queen Anne doll standing near a William and Mary court cupboard. Each room measures approximately 3ft × 4ft (0.9m × 1.2m).

In this survey of dolls' house furniture I have neglected the pieces that were made after World War II – largely because they are not to my own taste. The fashion for full-size furniture with spiky legs and plastic, imitation leather upholstery was emulated by some manufacturers of dolls' house furniture. Some companies continued to make "period" and "Jacobean" miniature items, but it is possible to find dolls' house furniture with "triple" plywood around the sides. In most instances, the plywood used was too thick for the purpose. Now that reproduction furniture seems to be back in favour for real-size pieces, there is also a wide range of dolls' house furniture in Gothic, Victorian and Edwardian styles available in various degrees of quality and priced accordingly.

OPPOSITE:
The study of the Nostell Priory baby house contains a burr walnut bureau/bookcase, an exact miniature of a full-size piece. This highly polished piece is 10in (25cm) high and 6in (15cm) wide. It has two glass doors covering two long bookshelves, and the fall front of the bureau opens to reveal pigeon holes and a red leather insert.

Also in the study is a games table, made of walnut, which is 3¼in (8cm) high and 4¾in (12cm) square. Other items of cabinet-made furniture in this baby house are, in the red bedchamber, a walnut chest of drawers, all with brass handles and escutcheons, and in the room on the top floor, to the left of the yellow bedchamber, an interesting hand-embroidered pole screen, a walnut knee-hole desk, with a centre drawer and four small drawers at each side of the knee hole. The easy chair in this room is 5½in (14cm) high and 3½in (9cm) across. *Courtesy: Lord St Oswald and the National Trust; photograph: Frank Newbould*

PRECEDING PAGE:
The bedhangings, curtains and chair seats in the chintz bedroom of the Nostell Priory baby house are all made from household chintz. The bed linen is sewn with the most minute stitches, and the draped dressing table and doll basket are home made. The dressing glass is cabinet made. Both dolls are wax: the nurse is 7¼in (18.5cm) high and the child is 4in (10cm) high. Both are beautifully made and clothed.

In the red bedchamber (see page 11), the lady of the house stands near a silver tripod wash bowl and stand. Also made of wax, the doll is 8½in (22cm) tall and wears a red and cream, silk robe that is open to reveal a white lace-trimmed undergarment with a deep flounce of the same material as the robe. The doll has glass eyes, painted features, a stuffed body, wax lower limbs, including wax boots moulded in one with the legs. The hair is real and is covered by a white lace-trimmed cap with lappets. The large lace fichu around the doll's shoulders reaches to the waist and there are generous falls of lace from the elbow-length sleeves. In the room to the left of the dining hall is a doll of similar construction, 7¼in (18.5cm) high, wearing white cambric and a gown covered with small pink sprigs.
Courtesy: Lord St Oswald and the National Trust; photograph: Frank Newbould

RIGHT, CENTRE:
This wonderful collection of gilt dolls' house furniture and accessories is believed to be French or American. This style of furniture with an "ormolu-type" finish is rapidly becoming known as Tiffany furniture after the similar items in the Tiffany-Platt house, in the Washington Dolls' House and Toy Museum. The items illustrated were made at the end of the 19th and the beginning of the 20th century, and, as an indication of scale, the dresser is 6in (15cm) high. Similar items were made in England, France and Germany. *Courtesy: Sotheby's, London*

RIGHT, BELOW:
Made at the end of the 19th century or during the early years of the 20th century, this suite of fretwork furniture is French. Similar, but rather heavier looking, items were also made in England, America and Germany.
Courtesy: Christie's, London; photograph: A.C. Cooper

FROM THE *ILLUSTRATED LONDON NEWS*, 1851

Messrs. Wittich, Kemmel & Co. of Geislingen have an upright case crowded with bone and ivory toys for children. The taste and finish of these are worthy of admiration. The Swiss cottages are in excellent proportion, admirably carved. The dolls' furniture, cut with the nicest detail. An occasional stain to imitate horn, giving a variety to the colour of ivory. The miniature sideboards, the cabinets, dumb waiters with their champagne glasses, hot water jugs, bottles, etc. are cut with a sharpness equal to many of the more pretentious examples of the Chinese. The baby's rattles and whistles are of novel design, the puppets looking-glasses [lorgnettes?] – open-through work are singularly tasteful for this description of toy. Every attempt, however, to turn the human figure, is a signal failure.

J. Rominger, Stuttgart: Tin and glass toys, lamps, decanters, glasses and every conceivable vessel, in little, to teach the young idea how to drink. In a second case is dolls' furniture in bronze, which exhibit some taste, and cast metal baskets of open work, requiring very little attention to make them really good.

OPPOSITE, ABOVE:
Dominating this illustration of dolls' house items is a rare set of furniture upholstered in orange cotton on which is printed, in black, a design of a cornucopia with garlands of leaves, flowers and a butterfly; the upholstery is edged with a trim of gilt paper. The *chaise longue* has an elaborately shaped back, a rather uncomfortable-looking bolster and a shaped foot rest. The two chairs are part of a set of five and all are decorated in a similar style to the *chaise longue*.

On the left is a metal upright piano with a matching stool. The soft metal workbasket on a two-legged support with four feet contains purses, a paper knife and balls of silk. The sewing machine is also of soft metal; it has a foot treadle. Next to the sewing machine is a filigree decorated metal pelmet and curtain rail. The cast metal table, which is of French design, has an inlay of simulated leather. Next to that is a galleried wine or silver table of two tiers, and, in front of the table, is a soft metal decanter holder with five glasses (sadly, the decanter is missing). The painted miniature perambulator and its occupant are of a slightly later date than the furniture, which was made between 1870 and *c.*1895. All the items are German made, although some are to French designs.
Courtesy: Christie's, South Kensington; photograph: A.C. Cooper

These bone and ivory items were made in Germany during the 19th century. The carved and turned *étagère* of three shelves displays a variety of miniature items. The finely carved and pierced piano, supported on cabriole legs, is surmounted by a three-shelved bookcase enclosed by double doors; it is 6¼in (16cm) high. The mirrored dresser with book or display shelves above a double-doored cupboard is 5¼in (13cm) high. The fine upright piano at bottom left is 2¼in (5.5cm) high. Next to it is a carved bone chess set on a table 2½in (6cm) high. In front of the 3¼in (8.5cm) high *secrétaire* is a tray and coffee set, while another *étagère*, 3½in (9cm) high, bears a display of bottles and ewers.
Courtesy: Phillips, London

5
DOLLS' HOUSE DOLLS

Exterior of Tiffany-Plat House.
OPPOSITE:
Date Mid 19th century
Maker Unknown American
Height 52in (132.5 cm)
This elegant sandstone-coloured mansion, albeit a child's toy with wide steps to provide seating, is one of the many remarkable houses on view in the Washington Dolls' Houses and Toy Museum and is features in Dolls' Houses in America by Flora Gill Jacobs. This is one of my favourite houses and I must draw the reader's attention to the superb 'Tiffany' gilt furniture in the large drawing room. So many choice and rare miniature items tastefully arranged throughout the house, combine with the ten dolls and baby to make the house come alive. Look again and again and revel in the delights of this charming tranquil abode.
Courtesy: Flora Gill Jacobs and Washington Dolls' Houses and Toy Museum.

Over the years increasing numbers of miniature dolls were created. Designed as inhabitants for baby houses, dolls' rooms, kitchens and shops, these dolls had heads and limbs formed from wax, wood, cloth, leather, clay or terracotta, types of papier mâché, ivory, precious metals (for those destined for cabinet houses) and last, but far from least, gum tragacanth. Often, heads and limbs described as papier mâché, as a type of composition or even as carved wood were, in fact, made from gum tragacanth.

More dolls were created from the various materials in Britain than is generally realized, and they were often superior in both construction and clothing to some of their continental counterparts. During the late 18th century and 19th century Grödnertals and china-head dolls, dolls of Parian, bonnet-head dolls and bisque dolls were shown at the exhibitions that were held with increasing frequency throughout Europe, the United States and Britain. The numbers swelled in the 20th century. The selection of makers (see pages 116–31) barely scratches the surface of the manufacturers producing dolls at this time. Dolls' house families, sold in boxed sets, were manufactured as well, and these were, of course, more expensive. Soldier dolls were costly, too. The inhabitants of dolls' houses ranged from babies, children, parents and grandparents, to soldier visitors, grand, tastefully gowned ladies and be-whiskered gentlemen, as well as butlers, maidservants, nursemaids, housekeepers and cooks. All of these needed equipment, and novelty items, some from Christmas crackers, some souvenirs from fairs and from the seaside, were added to swell their finery.

Dating a dolls' house by judging the age of the doll can be a thankless task, as so many changes have no doubt been made to the dolls' house families themselves and to their attire and status. Certain houses can be dated from the dolls, and examples are illustrated in this book – the Wyreside Park house on page 69 is a particularly good example. Always allow a tolerance of anything from ten to even thirty years when you are trying to ascertain the age of the doll.

You must learn to differentiate between genuine old bisque dolls and reproductions made today. The old dolls were made from a more "grainy" type of bisque, and, when they are examined under a magnifying glass, many imperfections will be revealed. Similarly with old china-head (glazed bisque) dolls. If you study a reproduction doll you will immediately notice the smoothness of the bisque and the high gloss on the china head. If you must mix these dolls with the genuine old dolls, try to show back views only so that the contrast between them is not too sharp. But, yet again, the choice is yours: the house and contents are yours and you must make up your own mind.

Many present-day collectors prefer reproduction dolls, and they are also turning their attentions to reproduction baby houses and dolls' houses. I have included illustrations of two of such houses: Britannia house (pages 48 and 49) is at the top end of the market, and the attractive Mayfield house (page 45), which was created by one person. All the dolls in Mayfield house are reproduction, and it is the type of house finding favour at the affordable end of the spectrum. I was most impressed by it, even though I do not normally favour reproductions myself.

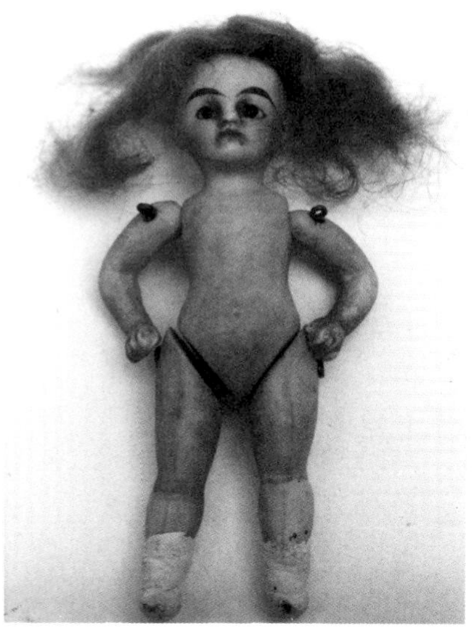

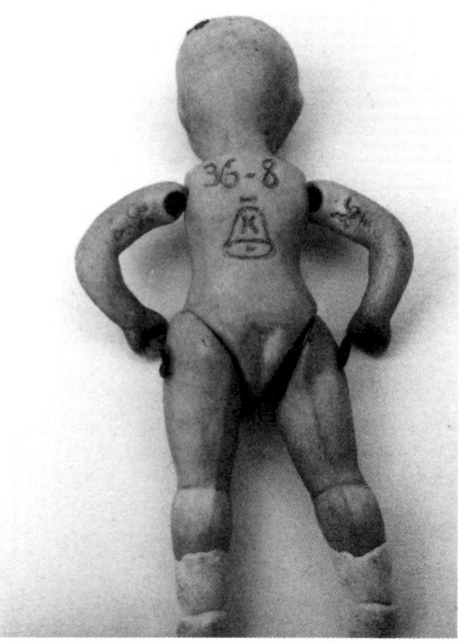

Date **1880s/1900s**
Maker **C.F. Kling & Co.**
Marks ***36-8 K* [in bell] on back of torso; limbs incised *36-8***
Height **3½in (9cm)**
This all-bisque dolls' house doll has a blonde mohair wig, fixed brown glass eyes and a closed mouth. The head and torso are in one, and the arms and legs are attached to the body by means of wire. C.F. Kling & Co. of Ohrdruf was founded in 1834 and continued to produce dolls until the 1930s. *Author's collection; photograph: Fabian*

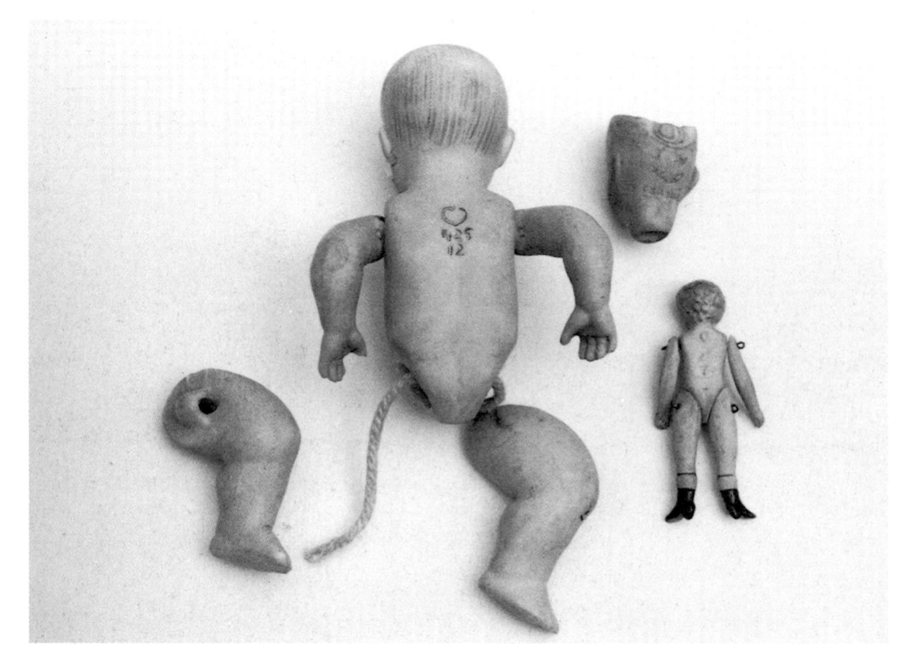

ABOVE, RIGHT:
These French *mignonnettes* (dolls' house dolls) have bisque heads, glass eyes, painted features and closed mouths. The cut-away crowns of their heads are known as "French slopes". They have composition bodies with painted shoes and socks. The wigs are mohair. Dating from the late 19th to the early 20th century, they range in size from 4in (10cm) to approximately 5in (14cm).

The tiny poupard held by the doll in the chair is a mere 2in (5cm) long; the bisque head is incised *3/0*, and the handle is bone or ivory. The head has been removed from the unclothed body to indicate the way in which the limbs are attached. The arms are secured by thin elastic, threaded through the body and through a hole at the top of each arm, and secured by a knot. Each leg has a separate length of elastic, knotted through holes at the top of the thighs, threaded up the body and attached to a small wooden bead in the socket of the head. An alternative method is to use wire when you are restringing. Bring each length of wire up through the body and the head, turning the wire over the rim of the crown at the sides of the head; re-attach the wig to hide the wires.

The French chairs are of unpainted wood. The side chair on the left is upholstered in pale blue silk with silk fringing; the two *fauteuils* (armchairs) are covered in pale blue quilted silk and have gilt paper decorative trimmings.

The four dolls have incised size numbers on the back of their heads, but these are too faint to decipher.

Author's collection; photograph: Fabian

Date **1920s**
Maker **Bruno Schmidt**
Marks **[heart] *425 12* incised on back of torso**
Height **5¼in (13cm)**

An all-bisque baby doll with a painted bald head and painted facial features. The bent limbs are attached to the body by means of elastic threaded through the torso and through the bisque "loops" that are integral to the limbs. The bisque loops break very, very easily, so take the greatest care if you ever have to restring such a doll. Bruno Schmidt of Waltershausen was founded in 1900, and the factory made dolls and babies of celluloid and of wood as well as of bisque. The "heart" trademark was registered in 1908.

The smaller doll is by Kling; it bears an incised *O* and the bell trademark. The doll is 2¼in (5.5cm) tall, but it appears to be much smaller than the Schmidt doll, which has a rotund body and appears to be much larger.

The doll's head, which is incised *J.* [anchor] *V.* and *France*, is by J. Verlingue, a company based in Montreuil-sous-Bois. It is almost 1in (2.5cm) long and dates from the early 1920s.

Author's collection; photograph: Fabian

This glass-fronted box which is 13in (33cm) wide, 18½in (47cm) high and 2½in (6cm) deep, contains as many examples as possible of the type of dolls' house dolls that might appear for sale and that would not be too expensive. The dolls date from c. 1830 to 1930; there are some early wooden dolls, "bathing" dolls, "frozen" Charlottes, a china-head doll with a sawdust-stuffed, calico body and some all-bisque dolls. They are of mainly French and German manufacture.

The chair at the left-hand side of the top row is made from feathers; it dates from c. 1840 and is 4½in (11cm) high. The china-head doll, third from the right, third row down, was awarded to me by Mrs Graham Greene as a prize in a competition she had organized. Invited to suggest how to dress a small doll, I proposed using men's discarded small-patterned silk ties.

The chubby-bodied bisque bathing doll, second from left in the third row, dates from 1900–20. Although unmarked, it may be attributed to Kestner – the painting, the clear-cut features, the moulding of the ears, the dimples in the arms and knees and the texture of the bisque are of the quality one would associate with that manufacturer.

The 1920s French bisque-head doll, third from left on the bottom row, is 5in (13cm) tall. The features are quite well painted and the open mouth shows tiny teeth. The maker of the dolls is clearly shown on the back of the head, which is marked *Paris, Unis France* [in an oval] and, underneath, what appears to be *1–30* and, underneath that, *o*.

Author's collection; photograph: Fabian

ABOVE:
Date **1900–25**
Maker **A. Bucherer**
Marks ***Made in Switzerland Patents applied for* impressed on torso**
Height **6½in (17cm)**
This large dolls' house doll of patented construction has a composition head, hands and feet and an aluminium torso. The limbs are connected to the body by 13 ingenious swivelling ball joints made of tin. A. Bucherer, of Amriswil, Switzerland, manufactured a variety of dolls during the 1920s, including some comic-strip characters.
Author's collection; photograph: Fabian

Many dolls were home-made from ideas and patterns in children's books, magazines and, of course, in Mrs Beeton's *Book of Needlework*. The little seamstresses did not work to fine limits, and one doll could vary widely from another made from the same pattern. Novelty dolls find homes either inside a dolls' house as a character servant or on the doorstep as a miniature pedlar woman, complete with miniscule basket and even tinier wares, or as a male hawker or tinker. Dolls that have been made up as pen-wipers are sometimes found with best quality china heads and exquisitely beaded outer-garments, the flannel wiper hidden beneath the dress. They will stand up un-aided because of the thick, stiff wipers underneath. I saw an amusing pen-wiper doll in a glass case at the Essex Museum, Salem, Massachusetts. It stood unaided, 3½in (9cm) high, was well dressed, and, had it not been for the details given on an attached card, I would not have realized what its unusual construction was. The card read: "Once I was a wish-bone and grew upon a hen, now I am a spinster made to wipe your pen." The doll had a wax face mounted on a chicken's wish-bone "body".

Many a good home-made doll has been constructed from a nut on a body made of a pipecleaner or a piece of rolled-up linen. The wrinkles of the dried-up nut resemble an aged person. Cut down to size, wooden

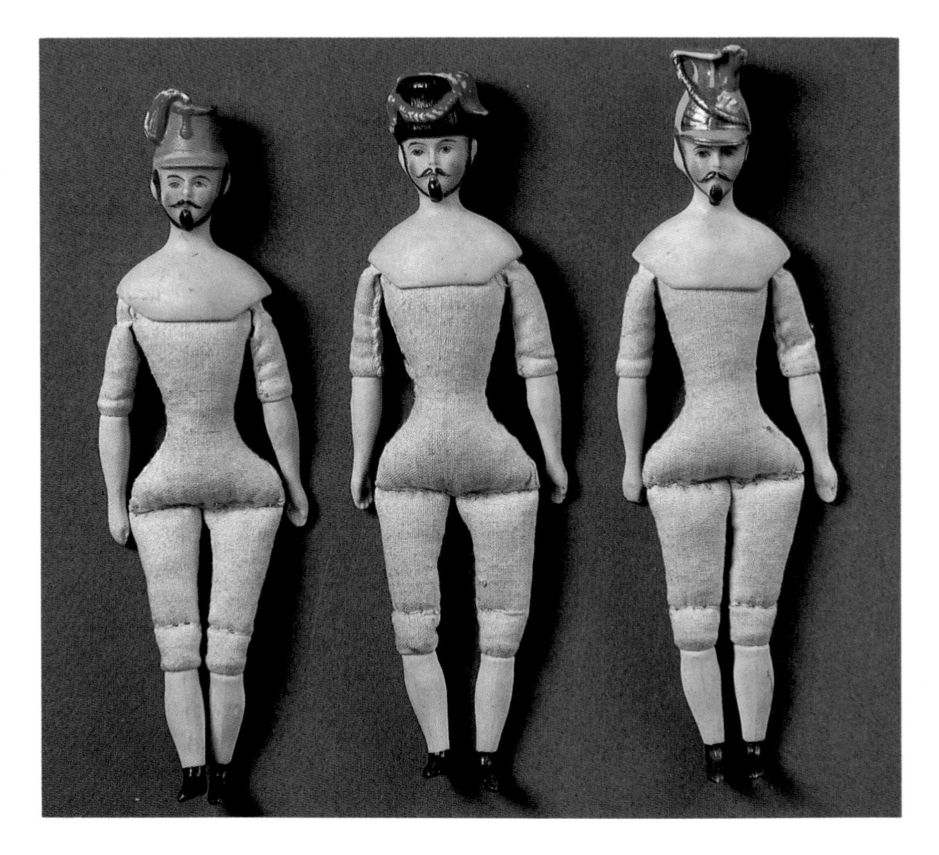

This trio of bisque shoulder-head dolls' house dolls is displayed in the Abbey House Museum, Kirkstall. The elaborate military headgear is moulded, and the dolls are attributed to C.F. Kling & Co. The lower arms and legs are bisque, and the bodies are pink calico, stuffed with sawdust. This type of doll is found in sizes ranging from approximately 4½in (11cm) to more than 10in (25cm) tall.
Courtesy: Abbey House Museum, Kirkstall, Leeds; photograph: Frank Newbould

clothes pegs can be made into chubby-bodied dolls. Your dolls' house dolls will look more human if their garments are puffed out a little here and there, and a cone-shaped stand specially made to fit round the waist underneath the skirt; an appropriate gauge of bound millinery wire is ideal material for this task. Aprons and removable fine cambric parts of their dress will benefit from a gentle wash, spray starched and "finger dried". Use your own common sense and do not shirk from these minor fiddly tasks – the end result more than repays the effort.

Dolls' house dolls which arrive at their new abode with no clothes at all often lie neglected for years awaiting attention. A small doll is more difficult to dress than a much larger doll. Extra fine needles, fine lawn and lightweight silks and tiny lace edgings and even smaller tapes and ribbons, tiny beads for use as buttons, pure silk or extra fine thread must be assembled. Books on costume must be studied and styles selected to suit the age of the doll and the position it will take in your dolls' house or baby house, if you are fortunate enough to possess such a treasure. Look at old engravings in books and use them to identify hair styles and complementary attire. They also sometimes contain items of furniture, making it possible to dress an unclothed doll in the correct style and to place furniture of the period alongside.

It is well documented that Princess (later Queen) Victoria dressed dolls to represent various characters – they have been illustrated in many books. Some were dressed by the princess herself, some by her governess, Baroness Lehzen. Illustrated here is one of the princess's Grödnertal dolls. It may, in fact, have been a discard as the legs below the knees are missing. It was sold at auction by Graves Son & Pilcher of Hove and Brighton on 20 September 1965 as part of the contents of a house. The auction was held by order of Lieutenant-Colonel A.H.S. Holden and the Executors of Mrs M.J.L. Holden, formerly lady-in-waiting to H.R.H., the late Princess Louise, Duchess of Argyll. The lot was described as "a pair of plaster and wood puppets . . . a miniature Victorian doll dressed by Queen Victoria". The purchaser allowed me to buy the doll from her when she learnt of my especial interest in wooden dolls. The doll dates from c.1831–3 and measures 4in (10cm) to the hem of the gown. Queen Victoria's dolls may be seen in the London Museum, and those who are particularly interested in them should read "Queen Victoria's Dolls" by Frances H. Low in volume VI of the *Strand Magazine* covering the period July to December 1892.
Author's collection; photograph: Frank Newbould

6
BUYING, DATING AND EVALUATING

Perhaps the most obvious question a novice collector will ask is: where can I find a dolls' house? There are several options open to you, but you should choose the method best suited to your lifestyle. You could respond to an advertisement in a specialist magazine or newspaper or place an advertisement yourself; or you could buy from a shop that specializes in selling dolls' houses and all that goes with them. With luck, you might even find a dolls' house in an antique shop that does not normally deal in such items. Go to doll, dolls' house and miniaturists' fairs, and look out for collectors' sales at auction houses, which will include dolls' houses, furniture, decorative items for dolls' houses and dolls' house dolls. Such auctions take place several times a year. If you go to a major city to view prior to the auction, seize the opportunity to visit specialist dolls' house shops, for nowadays, there are many excellent shops selling by retail or mail order. Guard against buying on impulse and plan before you buy. If viewed in a huge building or high-ceilinged room a large dolls' house may look deceptively small or medium sized. But when you get your purchase home, it may seem to have doubled in size. Always buy what you like, what you feel able to accommodate and what you can afford.

There are certain very practical considerations to take into account before buying. First, ascertain the size – height, width and depth – and whether the house is front, side or back opening. Second, decide where you would place the house in your own home and measure very carefully to make absolutely sure that you will be able to open the dolls' house once it is in position. This may seem an obvious precaution but you have to remember that if the façade opens in two parts, you will need the depth of the house and half the width plus "working space". If the house opens at the back you will need the space to walk round it.

If the dolls' house is large and some distance from your own home, transport could be a problem. The solution is to "be prepared". Make enquiries and try to get quotes from more than one transportation company in advance. If you buy at an auction, remember that at London auctions, if the dolls' house is not removed the same day, or the day after, it will be removed to a warehouse and you will be charged storage until it is removed. Many inexperienced as well as established collectors have been caught out over transportation. Not everyone has an estate car, and most collectors have a tale to tell of herculean efforts to get a dolls' house into a saloon car or onto a roof rack.

Are dolls' houses, their furniture, and their dolls good investments?

This is not an easy question to answer, if only because much depends on quality, condition, and rarity. As a general rule, if you buy and then wish to sell within a short time, you may well have to sell at a loss. If you are prepared to take a long-term view – perhaps as much as four or five or even ten years – you could make a profit. It will depend upon many factors. The established collectors whom you might have expected to tempt may have reached saturation point, so that they do not wish to add to their collections. Several collectors may have decided, independently but almost simultaneously, to sell items from their collections, so that there are already too many houses chasing too few purchasers. Buy what appeals to you and buy for your own pleasure, rather than with the hope of profit.

Remember the advice "buy in haste, repent at leisure", for unless you have had the extremely good fortune to find (and be able to afford) a fully equipped house, your problems may only just be starting. What happens if you buy or inherit or are given a relic that is in need of restoration or repair or that has been overpainted in unsuitable garish colours? What if your new house is in need of wall and floor coverings, or you cannot immediately afford expensive furniture, furnishings and dolls?

The advice I give in relevant chapters is based on my own experience and knowledge as a collector. You cannot buy experience, but you can gain it by using common sense. If you get a "gut" feeling that something is not as it should be, obey your instincts and look again and again. There are very few collectors who do not eventually gain a sort of "sixth sense",

The interior of the German red-roof house illustrated on page 137. The wallpapers are original, but the furniture is of various dates. The kitchen is actually a commercially made insert, dating probably from the late 1920s. Note the toilet to the left of the stairs and the remains of the rather clumsy home-made electrical lighting.
Author's collection; photograph: Frank Newbould

Date 1927
Maker Unknown English
Height 47in (119cm)
Width 54in (137cm)
Depth 31in (79cm)

One could be forgiven for thinking that the Surtees dolls' house was built in 1890 or 1900 rather than in 1927. It is an interesting house, and not just architecturally, for the furniture and furnishings are of varying periods, some commercially made, some made by craftsmen. The seven-legged dresser in the panelled dining room was fashioned by a master craftsman, as was the meticulously made half-open door. The smaller scale "oak" furniture in this room was made by Elgin of Enfield. In the room above is a Waltershausen washstand dating from c.1880, and, in the room across the galleried landing, a gentleman is taking his ease surrounded by some exceptionally well-made mahogany furniture with decorative inlays. In the parlour below home-made pieces blend with commercially made items. Note the miniature harp in the foreground. The attractive staircase is enhanced by the stained-glass window. The housemaid shoos away the cat while a candlestick telephone rests on what could be a Tri-ang oak chest.

An unusual feature of this house is that most of the occupants were made by the firm of A. Bucherer, a Swiss doll-making concern whose mark appears on a variety of metal-bodied dolls with composition heads. Bucherer made dolls to a patented design that allowed them to assume many different positions. (An unclothed Bucherer doll is illustrated on page 92 to show the many ball joints used in a doll of this type.)

Courtesy: Bowes Museum, Barnard Castle, County Durham

and, in any case, obey the maxim "when in doubt, do nothing".

You can buy knowledge by subscribing to auction catalogues and follow-up lists of prices realized. In this way, even though you may not be in the market for an immediate addition to your collection, you will be able to view and examine the houses, shops, furniture, dolls and so on. Observe what does come on the market, and which items appear frequently and which infrequently. Notice how prices rise or fall, depending on market conditions. The catalogues will, in time, become additions to your library of dolls' house books and will always be there for reference purposes. The knowledge will also give you confidence to enter a specialist dolls' house shop, which many beginner collectors are too timid to enter, to ask for full details or the price. If you are such a collector, your fears are groundless. Dolls' house dealers are friendly people and often go to endless trouble to assist a novice. They may, indeed, be prepared to "hold" a coveted house on payment of a suitable deposit and the promise to pay within a reasonable length of time.

INSPECTION AND ASSESSMENT

Begin with the exterior. Study the house carefully from all angles. Look for damage or missing parts. In 20th-century commercially made dolls' houses, for example, steps were often secured only by two nails or by glue that has chipped away, and they have become detached and lost. Look for nail holes, traces of glue, or slight roughnesses on the roof that reveal where a missing chimney once sat. Examine the painting, the glazing and the architectural details – nothing must be overlooked. Although it may be difficult, if it is at all possible examine the underside *with care*, looking for marks, writing or a label – the trademark of Christian Hacker has been found. Concentrate, for you are also looking for evidence of woodworm. Smell the wood, trying to detect any lingering smell of dampness or woodworm treatment. If you suspect that the house has been treated, look inside the house to see if the wall coverings have been stained. Finally, do not forget to examine the stand, if any. Is it contemporary or some sort of make-shift piece?

Turn then to the interior, checking the hinges and fastenings of the opening sections as you do so. Is the wallpaper original? If so, and if it is of pleasing design and has mellowed with the house, then you are fortunate. But if it clearly is an unsuitable later replacement, do not worry unduly: the original wall-covering may lie underneath and be revealed after stripping off the offending top layer. Examine the fixtures, if any exist.

Do not let two or three marvellous rooms excite you so much that they blind you to less obvious imperfections. A torch will be useful, to throw light into dark corners. A small mirror will enable you to view otherwise inaccessible parts of the house. A magnifying glass will let you scrutinize doubtful areas. A torch will be useful, too, for large dolls' houses are often thrust out of the way into ill-lit rooms. Be critical. The good points of the house will speak for themselves. You are looking for snags and anything you could not live with. If the wall coverings are damaged, assess the difficulty of stripping them and of finding replacements in keeping with the age of the house. Examine the floors, which may be polished, covered with oilcloth, or painted in a tiled or parquet design, with the same doubts in mind. Survey every part of the house a second time (if circumstances permit) preferably on another visit. It is amazing how initial euphoria may collapse on a second inspection.

If the house is unfurnished, study the interior again. Stand well back and assess the proportions of the rooms in relation to the size of the house. Are the rooms narrow with high ceilings, wide with low ceilings or proportioned "just right" with a good "balance". Look, too, at the depth of the rooms. The measurements and proportions will govern the ratio and scale of the size of furniture needed to furnish the house. If, for

example, the room is narrow with high ceilings, it is difficult to judge the scale. Your eye is the best judge when it comes to deciding how miniature is miniature and when "miniature" ceases to be classified as dolls' house furniture and comes into the category of apprentice or cabinet-makers' pieces. Indeed, dolls' furniture may simply be named "over-size", a definition often used in dolls' house collectors' parlance.

If the house is furnished and, based on your study of different periods, styles and quality of furniture, artefacts and dolls' house dolls, you are satisfied that the contents are either original to the house or part original and part additions over the years, all that remains to be agreed is the price (plus the cost of transportation). The short answer to what is the right price is "what can you afford?" If the house is in a specialist shop or in private hands, you may have to make an immediate decision. You face a dilemma. If you decide to think it over you may risk losing the house, but it is better to be safe than sorry: a dolls' house can be a very large white elephant. As always, when in doubt do nothing. If you cannot get the house out of your mind, then it is for you, and, if necessary, you must organize your finances, possibly by parting with lesser items in your dolls' house collection.

If the house is coming up for auction, you should make a note of the contents. Make a rough estimate of the bric-a-brac and concentrate on the furniture, clocks, chandeliers if any and, finally, the dolls. The house may possess only oddments of furniture and few dolls or none at all. The contents may be of a much later period and most unsuitable for the house. In both circumstances, calculate the cost of furnishing the house in period, taking into account the likelihood of suitable plenishments becoming available. Another complication could be that the house is very much younger than the furniture, which could be inherited items, kept for sentimental reasons when the original house was otherwise disposed of. You must then estimate the cost of a house of the correct period.

DATING

For the purposes of the collector, dolls' houses can be said to fit into one of the following categories:

Early baby or cabinet houses, dating from 1600 to about 1790;

Baby or cabinet houses (sometimes Regency or Georgian) dating from 1790 to approximately 1830;

Antique or Victorian houses, dating from 1830 to about 1890 (if they are in the appropriate style, they may be known as baby houses);

Edwardian houses, dating from the period immediately prior to World War I;

Between the wars, furniture and dolls' houses were still produced by many English, European and American manufacturers in the Edwardian style;

Post-World War II houses, furniture, dolls and so on of contemporary design;
Craftsmen or reproduction creations.

The above list is not intended to be a *mélange* of terms and classifications; it is merely intended to offer some guidance for beginners and less experienced collectors. Because so many dolls' houses look older than they are, judging their true age can be a tricky business, even for experts. All the indications – types of wood used, methods of construction and architectural details – must be surveyed and the evidence obtained in the initial inspection collated.

Has anything been overlooked? We must look for clues in obvious places. Is there a lock on the house? If so and the house appears to be 18th century, check the lock and the hinges also for means of attachment to the wood, whether by hand-made nails or pins, or screws. If need be, extract a screw, if screws were used. The screws or nails used may help us. In England screws for attaching metal fittings to wood were practically unknown before the middle of the 17th century. When they first came into use, their threads were filed by hand and the screws did not taper to a point. Screws with gimlet-pointed heads seem to have appeared at the time of the Great Exhibition in 1851. None of these dates, though, should be treated too reverently – a conservative cabinet maker or joiner might have used old-fashioned screws for years after the new versions became available.

Paints and colourings provide other clues. Oil varnish was extensively used after, again, the middle of the 17th century, which was also about the time when China varnish (in which gum was the main ingredient) was introduced. If you wished to carry matters as far as chemical analysis, a scraping of the paint could provide a clue.

If the house you are assessing is furnished, and the furniture seems in a style generally contemporary with the house, then that will help you to date the dolls' house. That is one reason why you should study dolls' house furniture of as many styles and periods as possible.

You must pay particular attention to wallpapers and wall coverings. If they are wood, is there a chair rail, wainscoting or cornices, or are the walls painted or stencilled? It is necessary for you to study the techniques employed from the 17th century until well into the 19th century. In the late 17th century silk and damask were fixed directly to the walls, while découpage was used on suitable papers. Celia Fiennes, referring to Newby Hall in Yorkshire in 1697, notes that "the best roome was painted just like marble". Painters and decorators of the time were trained to paint not just walls, but pilasters, panels, murals, ceilings, cornices, fireplace surrounds and items of furniture to simulate marble and tortoiseshell. They were skilled in the art of feathering and stencilling above wainscots; distempering was popular, too because of its cheapness. During the reign of William and Mary (1689–1702) a duty of 1d (penny) per square yard had to be paid on plain paper. There was an increase on this burden

Date **c.1675**
Maker **Unknown English**
Height **50in (127cm) (including stand)**
Width **55½in (141cm)**
Depth **22½in (57cm)**

This extremely fine and rare oak baby house affords a unique glimpse into the past. Each of the two floors has two glazed doors of six panes each. There are spiral-turned columns above a false-panelled central front door, which is flanked by fluted pilasters, and there are two panelled pilasters at each end. The front door opens to reveal one room on the ground floor and two above, divided by the original partition with arched doorway. It was originally further divided into four rooms, as shown by rebates for the original partitions, with each room having a corner fireplace and chimney breast. The lower division was placed behind the front door (which was never intended to be opened). The sides still have their original gilt bronze carrying handles of baluster bale form, but there are several minor repairs – some shrinkage of the rear planks has been filled with oak fillets, and some nails on hinges have been replaced by screws. The stand, which has shaped rails on turned baluster feet, has been strengthened with blocks behind the main joints.

This is the early type of baby house that still reflects its origins as a piece of furniture although with the added architectural detail that later became seen in the true miniature houses. This wonderful and exquisite house was the earliest baby house to have appeared at auction this century.

Courtesy: Sotheby's, London

Date c.1860
Maker **Unknown English**
Height **49in (124cm)**
Width **59in (150cm)**
Depth **15in (38cm) (including base)**

Horatia's house was owned by Lady Ann Elizabeth Jones, who died in 1890, and later by her daughter, Lady Horatia Jones (an only child), who died in 1969 aged 93. Horatia's father was Sir Horace Jones, architect and surveyor of the City of London. He designed and executed many buildings in conjunction with Sir John Wolfe-Barry and planned a bascule bridge to be erected across the Thames, a project that was completed after his death. He was knighted in July 1886 and died in 1887.

Horatia's house, which is now in the Kansas City Toy and Miniature Museum, has all its original wall coverings and floor papers. The needlepoint carpets were made by Horatia as a child. Apart from the dolls, the house has all its original furniture and furnishings in its 12 rooms.

The nursery of Horatia's house (centre). The glowing red walls provide a startling background for the interesting scraps. Note the wooden baby walker, the simple metal beds and the red and white treen ware set.

The dining room (left) in Horatia's house contains a superb renaissance-style Biedermeier sideboard and fine-quality Waltershausen red leather upholstered chairs and a settee, together with a Victorian military gentleman.

Courtesy: Mrs Barbara Marshall and Mrs Mary Harris Francis, co-founders of the Kansas City Toy and Miniature Museum

during the reign of Queen Anne, when, in 1712, another ½d was added to the impost on "paper, printed, painted or stained". This made wallpaper an expensive luxury, and much was sought in Europe when owners of stately homes went on a "grand tour". Further increases in this form of duty occurred over the years until the tax was finally abolished about the middle of the 19th century.

Study existing wallpapers and other coverings carefully and try to judge if the paper is hand blocked. Try to establish if it was made before the days of machine printing. If the walls are not papered but stencilled over distemper, it may provide a clue in helping you deduce if the dolls' house was made in the first half of the 19th century – houses were made in 18th- and early 19th-century styles until well into the second half of the 19th century – or, as you hope, much earlier, possibly even in the mid-18th century. There are so many points to take into account when one is trying to arrive at a fair approximation of dates. Walls above wainscots were covered in tooled leather during the period under discussion, and if you have to restore a drawing room or library in an 18th-century baby house use the leather binding from unwanted old books. Wall coverings do provide useful evidence of the age of a house, but it is only part of the evidence; there are so many other factors to take into account. Nevertheless, if your dolls' house wallpaper is hand blocked it may indicate that the house was made before the days of machine-printed real-life wallpapers. If the walls are stencilled not over paper but over distemper it may mean a dolls' house of the second half of the 18th century to the first half of the 19th century. Many developments occurred after the Great Exhibition and cheap machine-printed wallpapers became freely available.

The architectural features of the house should, nonetheless, provide most information about its age. Look at, for instance, the window glass. If it is riddled with slight imperfections, it may well confirm that the house is 18th or early 19th century. In 1697 a window tax (duty) had to be paid on properties (with certain exceptions – dairies, some government buildings, for example). Country houses and middle-class homes bore the brunt of the imposition, by which a property with more than six windows had to pay a window levy, which was increased many times between the middle of the 18th century, reduced in 1823 and finally abolished in the middle of the 19th century. Take a look at dolls' houses in museums and possibly in your own collection. Are there any false windows – skilfully painted on to match the "real" windows? If so, the history of the English house may offer a clue to dating the miniature house. Blank windows to give symmetry to a design came in only after 1697, when a window tax was first imposed in England. The tax was frequently reimposed until it was finally abolished in 1851, the year of the Great Exhibition. It may, therefore, be possible to relate a black-painted window on a dolls' house to a particular period of window taxation. Stanhope House (page 109), incidentally, has black-painted imitation glass in fanlights. It seems that no knowledge is wasted when estimating the age of a dolls' house.

EVALUATING

The following is general guidance on the term "all original" and on what is likely to be found in dolls' houses made between the 18th and 20th centuries, up to approximately 1939.

In the 18th century, the structure of baby houses was the creation of individual craftsmen. A genuine baby house will have a suitably "weathered" façade, and will not have been repainted. The window glass will be intact, and the interior will have floor and wall coverings as first applied, although they may show signs of wear and tear. The fabric of any furnishing will be of the period, and the furniture will be miniature versions of the full-sized pieces of the day. They will be good quality and of polished wood; walnut was very fashionable. If the baby house is very large, all the furniture may have been specially commissioned for that house; such furniture is often described as "apprentice pieces". All decorative items, whether of silver, porcelain, gilt metal or glass, should not be later than the date of the house. Dolls should have wax heads and limbs and bodies of calico, kid-covered wood, papier mâché or tragacanth; there may be all-wood dolls, with painted features, or dolls with papier mâché or tragacanth heads. All should be wearing contemporary attire.

Such 18th-century baby houses are also sometimes classified as "all original with later additions". This apparent contradiction in terms is sometimes used when houses are offered for sale, and it is most frequently used to describe baby houses of individual character made between 1800 and 1840. The overall structure is as for an "all original" baby house, but commercially made furniture and wooden Grödnertal and Biedermeier and china-headed dolls will have been introduced.

Houses made from 1840 to the turn of the century are now known as dolls' houses. They are seldom found "all original", for, to qualify for that description, they should be as "first made", whether commercially or by an individual. The wall and floor coverings should be as first attached, and the furniture, furnishings, decorative items and dolls as assembled by the first, original owner of the house. Use your common sense here. The first child owner may have inherited some earlier pieces of furniture and placed these in the house alongside newer, commercially made items. Pay particular attention to the hair styles of any dolls in the house, and many items were made in the home following the instructions from books by various writers such as Mrs Beeton.

Houses made from c.1900 to just before World War II were commercially produced in Britain and Germany in large numbers. Houses were made, too, from plans given in trade papers, newspapers, magazines and so on. Furniture and furnishings were produced in the same style over relatively long periods with only slight variations, and it is difficult to define "all original" for this period. The furniture could range from 19th-century "stock" items, to home-made pieces, to Elgin, Lines Triang or Barton items. The American Tootsie-toy furniture was also imported into Britain. Dolls varied from "penny woodens", to low-browed china-head dolls, to, among others, the "Grecon" dolls made by Grete Cohn.

7
REPAIR AND RESTORATION

The golden rule of dolls' house restoration is to leave well alone. The most desirable condition of all is the original. Restoration detracts from the value of a house both aesthetically and financially. So do not intervene unless it is absolutely necessary. It is, however, permissible and indeed advisable to remove any ill-considered renovations and acts of desecration that have been committed upon an old dolls' house or its contents by over-enthusiastic renovators in the past.

THE EXTERIOR

The first task after getting your newly acquired dolls' house home is to give it a thorough scrutiny. Your pre-purchase inspection (see page 96) may already have disclosed some faults and damage, but now you can make a more leisurely examination.

Pay particular attention to the back of the house. This is often made of much softer wood than the main structure and is more liable to split and warp. If a split is not too wide it may be filled with a proprietary wood filler. Wider gaps may be filled with a wooden fillet. I suggest that you make and try out a cardboard template first and only when that fits snugly use it as a pattern to cut the wood.

Inspect next the underneath of the house. (You will almost certainly need help to raise and, carefully, tilt it.) It is possible that you will find writing, marks, or even a label. If so, consider yourself fortunate. Such finds are rare. More practically, any splits you find may be repaired in the same way as those in the back. If you find signs of active woodworm, you may elect to treat it yourself with one of the several proprietary woodworm killers on the market. Remember, though, that the preparation may work its way through the wood and stain the internal wall coverings. If the house is old and valuable, it would be safer to have the house professionally fumigated.

Turn now to the façade, sides and roof. The rule of non-interference still applies. If you have a lovely old house with only minor faults and disfigurements try to camouflage them rather than repair them. Cover slight damage to the façade by placing a male doll dressed as a gardener, perhaps pushing a wheelbarrow, in front of it. If there is a broken window, have a doll dressed as a workman standing on a ladder – he is repairing the damage. Or have a maidservant doll leaning out, shaking a duster, to hide the broken pane.

The dining room of Stanhope house (see also page 109) contains an extremely rare Waltershausen simulated marquetry suite. Most Waltershausen furniture is of simulated rosewood with either gold stencilled decoration or "ormolu" mounts. There were several makers of fine-quality dolls' house furniture in the Waltershausen region, and this particular set cannot be firmly attributed to one producer. The *escritoire* has a mirror back and is lined with green paper; there is one drawer above and two below the fall front, and it is 6¾in (17.5cm) high to the top of the pediment, 4in (10cm) wide and 2in (5cm) deep. The display cabinet has two shelves with gold edging; the whole of the interior is lined with the familiar green paper, and it is of similar dimensions. The square table is 2¾in (7cm) high with a top 3½in (9cm) square; the smaller side table is 2¾in (7cm) high and 3¼in (8.5cm) in diameter. The chairs, 4in (10cm) high with seats 2 × 1½in (5 × 4cm), are upholstered in dark wine coloured velvet with a gold edging. There is also a matching pedimented and scrolled wall mirror, 4in (10cm) long and 2¾in (7cm) wide. Green or blue lining paper is a feature of good quality Waltershausen furniture.

Author's collection; photograph: Frank Newbould

Sometimes, of course, restoration is essential. Many fine and rare old houses have suffered the fate of being repainted – often by well-meaning parents trying to make an old house look newer, fresher, and more appealing to a young child. Stanhope House in my own collection had suffered just such a fate. It had been painted a sacrilegious pale blue. All the toil and anxiety of removing the offending paint seemed worthwhile when the original "stone" quoins and faded rose "brickwork" reappeared (see page 109).

It is hard to give firm instructions about how to strip the paint in such circumstances, mainly because so many different types of paint have been in use over the whole period from early baby houses to modern dolls' houses. The only suggestion I can usefully make is that you should try one of the modern, gentle, proprietary paint removers or, if the paint is not too thick, sugar soap. Always try the treatment out first on a small test patch of paint in some hidden corner of the house. Do not use any coarse abrasives, such as sandpaper. I have had good results with the aid of a loofah, smoothing off with a pumice stone. Whatever stripping agent you use, take care to protect your eyes and hands and never work in a confined, unventilated space. Follow the manufacturer's instructions

When it comes to repainting, use a colour and a texture sympathetic to the period and style of the house. In the late 18th century and early 19th century many English houses were painted a pale fawn or honey colour and given a sandstone texture. Abbey Grange (see page 42) is an example.

Picture glass is ideal for replacing broken glass in dolls' house windows. Ordinary window glass is far too thick. Old picture glass is fairly easily obtainable and parts of the frames may be cut down for use as dados, skirting boards or cornices.

THE INTERIOR

Any newly bought dolls' house deserves first of all a good spring clean. Be methodical about this. Take out all the furniture, removable fittings, and dolls. A tea trolley alongside is useful to receive them and keep them in their groups, but failing this boxes and deep trays will do. Take great care to avoid damage to fragile parts of delicate furniture and, if accidents do happen, effect an immediate repair. For this you should have a non-drip gel-type glue beside you. Guard against damage to glass globes on chandeliers. Now remove any carpets and rugs, give them a shake, and brush them gently with a soft nail brush.

Dust the walls with a suitably sized soft brush – I use a set of cosmetic brushes solely for cleaning dolls' houses – brushing into nooks and crannies. Even the folds of curtains and dolls' dresses benefit from a gentle sweep down with a soft brush. Wallpaper should, at this stage, be treated gently. In Victorian and Edwardian times it was cleaned by rubbing with soft bread. This is worth trying, though present-day bread being what it is it may be better to use cotton wool.

Now you can see what more major tasks need to be undertaken.

MRS BEETON'S ADVICE ON MAKING A VICTORIAN DOLL'S HOUSE

Endless as is the variety of amusements to be found for the little ones, nothing gives so much real and lasting satisfaction as a doll's-house, and this, like many other things, can be made at home if there happen to be a good-natured big brother who will condescend to interest himself in the work. There are always packing-cases about, stored away in cellar or attic, one of which could be spared for the purpose; this, then, with a few deal boards, some two-inch screws, a pair of hinges, some nails and smaller screws, a hasp for the door, glue-pot, and last, but not least, the willing brother or uncle with his box of carpenter's tools, can be quickly converted into a charming *doll's-house*. The case, after being thoroughly cleaned, should be set on end, and the places for floors and partitions marked out, if only large enough to admit of two rooms, so much easier to make, as it will only want one shelf in the middle for the bedroom floor, the end of the case itself doing duty as a floor for the sitting-room. If large enough to admit of *four* rooms, then a piece of board should be sawn off evenly, the edges, all but the front one, smeared with glue, and this should be fitted into the case, at about the centre; this would be the bedroom floors; then, after proper measurement, another piece of wood should be prepared and slid in edgeways between this floor and the ceiling at about the middle; this will be the partition wall between the two bedrooms; for these should certainly both *be* bedrooms, not allowing one to be used as a drawing-room, for children may thus be taught, even in their play, that it is necessary to health and wellbeing that sleeping accommodation should not be in any way curtailed.

This floor and partition may be made firm by the use of the two-inch screws, which can be driven in from the outside, the heads being concealed by papering when the carpenters have completed their work, and the house is in the hands of the paperhangers. After this, a partition of the same kind will be required below to separate the sitting-room from the kitchen. The papering should be done before the door is put on, as the house is much easier to turn about then. White foolscap does best for the ceilings, and any scraps of wall-paper can be used for the other parts, only care must be taken that the *pattern* on the paper or papers is small, or the rooms will be dwarfed and ugly.

If the wax and china ladies who are to inhabit this little mansion are æsthetic in their tastes, and insist on a dado in their parlour, it can be made thus: – Take some white foolscap, such as that used for the ceilings, cut it to the length required for the walls, then with pencil and rule draw some faint lines on the paper perpendicularly and about an inch and a half apart. This done, cut some strips of coloured paper, blue, green, or red, whichever best suits the tone of the room, and paste these on the white paper (using the pencil lines as a guide), bringing them to within three inches of the bottom; then add a horizontal line of the same to hide the ends. Now comes the dado, the making of which will give intense delight and amusement to the little ones. If wanted very simple, turn out mother's collection of crests, pick out the darkest and arrange them as they look best on the white paper you have left below the horizontal strip of coloured. The crests should be stuck on with gum, as it is better for them than paste, and when tastefully arranged they have a pretty effect; but a much more elaborate one can be made thus: collect all the old valentines, Christmas cards, &c., those which you do not particularly care to keep, and pick off or cut out from them all the tiniest figures of birds, insects, &c., with which these works of art generally abound; these, with some bunches of flowers, miniature trees, and tiny cupids disporting themselves in the shade, will, if arranged so that the birds appear to be flying or perching on the branches of the trees, and the insects crawling about beneath among bright flowers and grasses, make quite a charming dado, but it is much more difficult to make than the other, though the planning of the attitudes for the different figures will, as we have before remarked, afford great amusement. In this, too, mother's paint-box will be required, as the foreground will want "touching up" a bit; but with a little patience and perseverance all slight difficulties can be surmounted . . .

The wallpapering finished, the door may now be put on, and to take off the box-like appearance it may be covered with folds of chintz firmly nailed on – unless there be any one among the party clever enough to cut away the wood and insert panels of glass; in that case the appearance would be much more elaborate, to say nothing of the pleasure for nimble little fingers in making up and arranging curtains to suit the different interiors. When the door is in its place the furnishing proper may begin.

FLOORS

If the floors are original, and in reasonable condition, whether painted or covered in simulated tile or parquet paper, leave them well alone. If they are damaged, get out your paint box, mix up matching colours by trial, error, and experiment, and touch up the design as necessary. Short-handled cosmetic brushes once again come into their own for working in a confined space.

As a generalization, the earlier the dolls' houses the fewer the carpets. Older houses have polished wooden floors and handmade mats and rugs in most of the rooms, tiled floors in hall and kitchen, and plain wooden stairs and landings. From the middle of the 19th century carpets appeared in drawing rooms and parlours and mats and rugs were scattered elsewhere. Many of these floor coverings were home-made improvisations, often by children. Children at school were, in Victorian times, often taught to make items for dolls' houses. Small woven mats still to be found in Victorian and even between-the-wars dolls' houses bear witness to these child labours.

Replacing stair carpets seems to pose problems for many first-time collectors, but is not, in fact, very difficult. Use braid of the period or silk or velvet (not nylon) ribbon, held in place by stair rods fashioned from whatever you can find, ideally brass. If you are sufficiently skilled, you may make a carpet of petit point. Whatever you use, the miniature carpet must not look glaringly new – it must seem to have been walked on. To achieve this effect you may have to fade or distress the fabric before laying it on the stairs.

Do bear in mind that only very recently have stair carpets covered the treads totally. Until the end of the 1930s they would have been simply a strip down the middle, leaving a space at either side of each tread that would have been stained and polished, stained and grained, or painted.

Date 1835–40
Maker Unknown English
Height 37½in (95cm) (to top of chimneys)
Width 29in (74cm)
Depth 19in (48cm)

Stanhope house, a delightful, front-opening English house, which originally had a lock, has four rooms. Much careful stripping away of the pale blue paint that had been added by an enthusiastic but misguided owner in the 1940s revealed the original painted simulated brick and stone quoining. The fanlight above the door has painted black panes. The conservatory-cum-greenhouse, which is believed to date from 1855–65, is English and contains a variety of brightly coloured plants.

Author's collection; photograph: Frank Newbould

WALLS

The golden rule, as always, is to preserve as much of the original wall covering as possible and to leave flaws well alone. A defective patch of wallpaper may sometimes be concealed by standing a piece of furniture in front of it or hanging a picture over it. If the period of the dolls' house permits, tapestry wall hangings can cover a multitude of faults. Think of the fun of the chase in tracking down fragments of genuine old tapestry of the right period.

Sometimes, though, you will be forced into taking remedial action. What to do, for example, if one or two walls of a room are unpapered while the others are completely covered in wallpaper in good condition? There are several options open to you. If it is a Victorian house, you can cover the walls with contemporary scraps (see Horatia's house, page 101). In dire cases, you can use the paper from the good walls to cover part of all four walls, creating a dado or frieze to make up the height.

If the walls have been repapered at some time with unsuitable – perhaps even modern vinyl – wall covering, you have no alternative but to try to remove it without damaging what may lie underneath. Moisten the paper, being careful not to saturate it, until you can gently peel or scrape it off. Fingernails are best for this task, but if you do not wish to ruin your hands then a stripping tool that will not dig into the paper beneath may be used. You may be lucky and find the original paper.

If, after all your efforts the underlying wall covering is beyond recovery, then you must exercise your flair and judgement in finding a contemporary wall covering that is both suitable for the room and sympathetic to the house. You should consider using tooled leather above dado or wainscot. Watered silks in a delicate shade of *eau-de-nil* or faded rose or a small-patterned brocade were used as far back as the days of baby houses. If you are a purist, you will want to wait until you find some original old paper or use a silk substitute. The choice is yours: the house is yours and you have become an interior decorator, albeit on a small scale.

When repapering a wall, it is best to cut a template from plain paper and make sure that this fits the wall shape. Only then cut the patterned paper. One difficulty with silk wall coverings is that the raw edges have to be turned back out of sight and their outlines may show up when the silk is pasted to the wall. A way around this problem is to fix the silk to a piece of card the same shape and size as the wall. You can then either fix this to the wall or create a free-standing section of three walls that will stand inside the room walls like a lining.

CURTAINS

Rich silks and velvets were used to curtain baby houses and early 19th-century dolls' houses. Simulated linen blinds came in towards the middle of the 19th century. Hand- and machine-made lace curtains were introduced, together with curtains of plain and patterned chintzes, damasks and brocades. Brush curtains gently to clean them, and, if you have to replace, try very hard to match material to period.

FROM AN APPLICATION FOR A PATENT, No. 294,981, DATED 11 JULY 1928, MADE BY MIGUEL ENRIQUE NEBEL, A URUGUAYAN LIVING IN PARIS, FOR IMPROVEMENTS IN FURNITURE ADAPTED TO BE BUILT UP AND TAKEN TO PIECES OR IN COLLAPSIBLE FURNITURE

This invention relates to furniture adapted to be collapsed or taken to pieces and reconstructed, such as tables, chairs, divans or other like articles and refers more particularly but not exclusively to toy furniture intended to be used as playthings by children, who will be able for instance, to construct from the different parts, miniature furniture for dolls, etc.

The invention refers more particularly to a new method of manufacturing or building up these articles of furniture, the method being characterized by the different parts intended to construct the articles of furniture being cut from a very thin plate or sheet in an appropriate manner, and bent so that slots, guides, projections and the like are formed for permitting of easy and rapid assembly of the parts. . . .

It is to be understood that the invention extends to the construction of other articles of furniture, especially miniature furniture, besides those described and illustrated.

In some cases the invention may be applied to the construction of steel furniture of ordinary dimensions.

I am aware that it has been proposed to form boxes for chocolates and sweetmeats with slots and tongues to enable them to be built up into toy furniture.

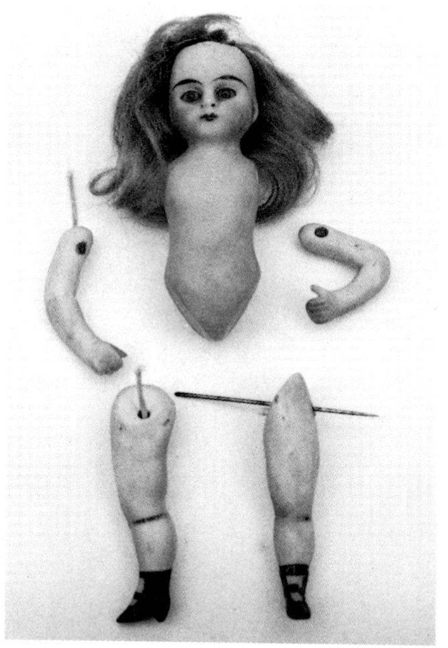

Date **1890s/1900s**
Maker **J.D. Kestner**
Marks ***4/0* incised on back of head**
Height **4in (10cm)**

This all-bisque dolls' house doll with a head and torso in one piece, has fixed blue glass eyes, a painted, closed mouth, a blonde mohair wig, and painted white socks and black strapped boots. The manufacturer, J.D. Kestner of Waltershausen, applied for a patent in 1894, D.R.G.M. No. 23 504, for china (bisque) "bathing dolls" (dolls' house dolls came under this generic heading). The application reads: "Porcelain bathing doll, closed at all sides, with holes through which the hollow arms and legs, which have massive pierced plugs, are connected by rubber thread, so that the arms and legs can be turned." The illustration shows the plugs in place with broken elastic attached. The plugs are made of wood and are very difficult to prise out if you have to restring because of the danger of flaking or even cracking the bisque. Care and patience are needed to pick out bits of the plug with a needle until the remains of the plug and the broken elastic come away. Wire is a practical alternative if restringing is necessary. See the illustration on page 89, which shows a Kling doll strung in this way.

Author's collection; photograph: Fabian

CEILINGS

If the ceiling is clear of any lamps or ornamentation and is badly stained or really filthy (and the house is not a rare 17th- or 18th-century baby house) either re-paper with an off-white or ivory paper or paint with a matt emulsion. Use white that has been "knocked down" by the judicious use of a touch of grey or pale fawn. This may take some time and patience, and you must try out the colour elsewhere (on a piece of board or card) and let it dry before proceeding.

Remember at all times when undertaking any form of repair, touching up (and this is only to be undertaken with great delicacy) or restoration, "make haste slowly". Do not restore any part of the house or its contents unless it is absolutely vital to preserve its structure. What is permissible, and indeed advisable, is to endeavour to remove the misguided attempts of over-enthusiastic renovators, who may have unwittingly performed acts of desecration on an old dolls' house and its contents. Be critical of your own workmanship after you have got the furniture, pictures, dolls and ornaments in position. Then and only then will you be able to decide if your efforts have been successful. If you are not satisfied, take up the challenge and start again. Achieving the standards you have set yourself will more than compensate for the effort involved.

8
CARE AND DISPLAY

Date c.1909
Maker Unknown English
Height (of case) 24in (61cm)
Width (of case) 22in (56cm)
Depth (of case) 11½in (29cm)
A fashionable Victorian and Edwardian pastime was to make a replica of the exterior of the family home and enclose it in a glass case. Mount Villa is one example of this; another may be seen, in miniature, inside the dolls' house illustrated on page 73, Norwich house.
Author's collection; photograph: Frank Newbould

You are the conservator of your dolls' house collection and you have a responsibility to ensure that the houses in your care are preserved from deterioration.

All dolls' houses need protection from extremes of temperature, from extremes of humidity and from strong light, especially sunlight. They can be seriously damaged in both the long and the short term if any of these factors are neglected. Central heating is perhaps the dolls' house's worst enemy; it dries out the wood and causes cracking and warping. Sunlight is harmful because it causes colours to fade and fabrics to rot. Do not place any dolls' houses close to a radiator or in strong sunlight near a window.

Ultraviolet light and heating can fade dyes and pigments and make papers and textiles brittle. Lithographed paper-covered houses and some older painted houses can also be seriously damaged if they are exposed to adverse conditions for any length of time. Most older finishes consist of natural gums (resins) dissolved in a solvent such as alcohol. Light energy breaks certain chemical bonds in these coatings, making them turn yellow or darken. They also shrink, crack and become less cohesive, and become more difficult to conserve safely. Wallpapers are also affected. Wood pulp contains lignin, a highly acidic natural binder for cellulose fibre of woody plants, which is particularly unstable when exposed to light. Almost all papers made since 1870 contain it.

Household pets can also cause harm. Cats seem particularly fascinated by dolls' houses – their curiosity makes them eager to get inside and investigate. Claw marks and other forms of damage can follow. Make sure that dolls' houses and cats are kept apart. Moths and other insects also attack dolls' houses. Some collectors advise putting camphor balls in the dolls' house interior, but the smell will permeate everything in the vicinity and you will never get rid of it. I find that cedar-wood shavings from a pencil, placed out of sight inside the house, are a less offensive but equally effective deterrent to moths and other insects.

Dust and dirt are enemies, too. If your collection is in one separate room it makes sense to cover each item with a dust cover made from an old cotton sheet or some other non-fluffy material. (Avoid plastics except for temporary cover when transporting a house.) The covers can be removed when you wish to enjoy or work on the collection or when you want to show it off to friends or fellow-collectors. Meanwhile, you will have saved the time you would otherwise have spent dusting.

MOUNT
VILLA
1905
H. BUXTON

FINDING MORE SPACE

Dolls' houses take up space – space that is often at a premium in a small house, flat or apartment. One way of making optimum use of the space available to you is to take to the walls. Room settings in glazed cases can be fixed to the walls like three-dimensional pictures. A shelf could be utilized to hold room settings custom made to size and shape. A disused picture frame could be turned into a shallow glass-fronted box to hold an elegant suite of miniature furniture and perhaps a doll or two.

Other space-saving ideas will occur to you. If you have a built-in wardrobe in a seldom-used room, you could construct in it your own version of a cabinet house. Normally you have to open a dolls' house to view the interior, and this is simply an extension of the same idea. I once carried the idea a stage further and accommodated four dolls' houses on a shelf running the full width of a large built-in wardrobe.

KEEPING RECORDS

Most dolls' houses are acquired without their provenance being known, but all dedicated collectors long to know the history and ownership of old dolls' houses in their possession. We cannot alter the past but we can make sure that we keep full records of the houses, contents, and dolls for which we have become responsible. The hope is that future owners will preserve and add to the "house-deeds-cum-inventory" thus begun.

For the house you should write down the date and place of acquisition, the name of the previous owner, the price paid, any of the house's history conveyed to you orally or in writing when you bought it, and a detailed survey of its condition at purchase, plus a "before" and "after" photograph if possible. To this you should add the history of the house while it is in your possession, beginning with any repairs or restorations you may have done and describing also any alterations or additions to the structure and any interior decorating you do.

You should also keep a record of the house's contents, beginning with an inventory of any contents at the time of purchase and going on to give details of any furniture or equipment you add – not forgetting any dolls' house dolls. Attach a small label to the back of the house itself noting the existence of your "house-deeds-cum-inventory".

DISPLAY

You will naturally want to be able to view your house's interior under the best possible conditions and you may be tempted to light an older house electrically. If you decide to electrify a house, do get a system especially designed for the purpose. These safe electrical systems are available from specialist shops and from specialist mail-order companies. Be certain, too, that you will obtain benefit from introducing lighting to an older house. It is not in a museum, being viewed for hours on end. You have only yourself to please, and a torch or lamp can be used to illuminate the interior long enough to show its delights to a visitor. The rule of making as few alterations as possible still applies.

FURNISHING THE HOUSE

I suggest here appropriate furnishings for houses of different periods. *Late 17th-century to early 18th-century houses* need few pieces of furniture. The floors, if original to the house, should be showpieces in themselves. Do no more than apply a water stain and wax polish to them. Similarly, the wood-panelled or wainscoted walls should be left alone, although tapestry wall hangings might be added if walls are in need of disguise. Apprentice pieces would make suitable furniture, depending upon the size of the house. Silver, porcelain and glass miniatures of the period would be appropriate additions. As for the inhabitants, wax, tragacanth, papier-mâché, and early Grödnertal dolls would all be desirable, as would early porcelain dolls on articulated wooden bodies.

For houses dating from the 1840s to the early 20th century a wide variety of furniture is both available and suitable. Choose from Waltershausen furniture, Rock & Graner tin (not everybody's choice, but among my favourites), and Victorian rosewood and simulated marquetry furniture. Furniture can also be extemporized. It is not difficult to create a four-poster bed and almost any small box will make a table if it is covered by a fringed tablecloth with a plant placed on top.

Dolls should be miniatures of real adults or children living in the era of the dolls' house. If you have to economize and introduce reproduction dolls try to position them so that they look as though they are doing something about the house and give them face-concealing hats or bonnets. To fit the doll to the period, see Chapter 5.

SETTING THE SCENE

One way of making a dolls' house come alive is to reproduce within it scenes of domestic life in miniature.

At Christmas, for example, set up a dolls' house for a family Christmas – a welcoming wreath on the front door, a tree with presents, stockings hanging from the feet of beds, a plum pudding on the kitchen table, in the dining room a table complete with crackers and a snowy white tablecloth. Miniature cards to stand on miniature mantelpieces are easy to make from cut-down real-life greetings cards. Remember to put red paper or cellophane in firegrates to give a semblance of winter warmth.

There are, of course, many occasions you could introduce to make your dolls' house become a home. A birthday celebration, complete with birthday cake, "jellies" on the table, could show "children", visitors from other houses, seated around a table. If you choose a wedding celebration, bear in mind that bridal gowns were not always white until about the middle of the 19th century.

Your aim should be to make your dolls' house live and become a home. A realistic scene should be like seeing times past through the wrong end of a telescope. If you look carefully at many of the illustrations in this book and in others listed in the Bibliography, you will see how often this has been successfully achieved.

MANUFACTURERS AND AGENTS

The following manufacturers and agents represent a selection only of the many companies and individuals active in this field. Manufacturers' marks are very occasionally found on the back of or underneath dolls' houses; suppliers' labels are sometimes found in the same places. Labels may be found on the boxes (on the lids or the sides) that contained furniture, which is itself very rarely marked on the front and sometimes underneath. The backs of dolls' heads may be marked; dolls may bear marks on their bodies, between the "shoulder blades", on the bottoms or on the soles of the feet; shoulder-head dolls (on kid or "nankeen" bodies may be marked at the front or back of the shoulder-plate or, very occasionally, on the soles of the feet. China tea services and so forth may be marked on the base; the boxes, if extant, may also be marked. Often, the maker's trademark and not the name appears on boxes, and manufacturers may have registered many marks and tradenames during the lifetimes of the companies. Advertisements in trade journals and periodicals often include the marks.

FRANCE

Banneville et Aulanier
Paris
19th to early 20th century
Metal furniture including gilt ("Tiffany") type and other styles; possibly "Orly" (or "Orley") type.
Trademark: *E.B.A.* [in a circle]

Bijard et Paillard
Paris
19th to early 20th century (successors to Borreau in 1895)
Some dolls' house dolls.

Blampoix (*jeune*)
26 rue Aumaire
Paris
from 1850
Dolls' house dolls and room settings. Blampoix *aîné* (senior) did not, as far as is known, make dolls' house dolls; he operated from a different address in Paris.

Borreau (*jeune*)
Paris
c.1875 to 20th century
Dolls' house dolls and accessories.

Botel et Soeur (also listed as **Borel**)
Paris
c.1850–c.1900
Wooden and papier mâché dolls' house dolls; some German furniture; dolls' house rooms and settings.

Bouchet, Adolphe H.
Paris
c.1875–c.1900
Dolls, including dolls' house dolls.
Trademark: *A-D Bouchet O* [in three lines]

Brasseur et Videlier
Paris
c.1850–c.1900
Dolls' house dolls.

Danyard
Paris
from 1850
Dolls' house dolls, 5in (14cm) high, but mainly dolls 10in (25cm) or more high; patented (in 1860) a bonnet-headed doll.

Delestaing
Passage de l'Opéra
Paris
c.1850–c.1900
Dolls' house dolls and room settings.

Doléac, L., et Cie
Paris
c.1880–c.1908
Dolls' house furniture and room settings; dolls' house dolls with both French and German heads.
Trademarks: *Paris L.*; *Paris L.D.* [on two lines]; *L.D.*

Fourot, Paul-Toussaint
Paris
end 19th to early 20th century
Dolls' rooms (including kitchens), shops, toys, games and knick-knacks.
Trademarks: *P.F.*; *Le Ménage Enfantin*; *P.F.* [intertwined]; *La Petite Ménagere*

LE MÉNAGE ENFANTIN

LA PETITE MÉNAGERE

P. F.

Gaujard, Alexandre
Paris
1850–69
Furniture and fittings; some dolls.

Gaultier, François
St Maurice, Charenton, Seine

F. 3.G

(Paris after 1900)
1860–1916 (possibly later)
Top quality dolls, including dolls' house dolls.
Trademarks: *F.G.* [in a cartouche with size number on opposite shoulder]

Gottschalk et Cie
Paris
c.1850–c.1900
Agent for many German manufacturers.

–M–
PARIS
GD
4

Grandjean
Paris
c.1850–c.1900
Dolls' house dolls.
Trademarks: *–M– Paris GD 4* [in four lines]; *Paris GD* [in two lines]; *GD*

Gratieux, Fernand
avenue des Moulineaux, Billancourt, Seine
Paris
late 19th century to 1906
Dolls' house rooms, furniture and ornaments, and dolls' house dolls.
Note the similarity of Gratieux's and Gaultier's marks.
Trademarks: *Mon Ménage*; *Paris Le Gracieux Déposé* [in three lines in an oval]; *Paris F-G Déposé* [in three lines in an oval]; *Tout Va Bien* [in three lines in an oval]; *Le Gracieux*

Gratieux, Fernand (*fils*)
14 rue Oberkampf
Paris
from 1907
As above

Guillard, Maison
Paris
c.1840–end of 19th century
Principally known for "Parisiennes" but also dolls' house dolls, room settings and furniture.
Trademark: *A.T. Guillard Jouets Rue Nve des Petits Champs* [in three lines in an oval; also in different forms on labels and stamped on dolls' bodies]

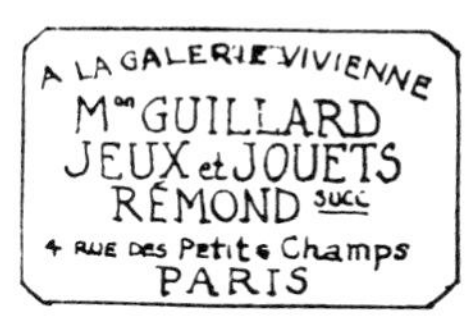

Hellé, André
Paris
from early 20th century
Dolls' house dolls, furniture and shops.
Trademark: stylized *ah* [in a circle]

Huret, Maison (Mme and Leopold)
Paris
1849–1920 (possibly later)
Superbly designed cradles and fine-quality furniture in wood or metal and various accessories. Obtained many patents for dolls, but, the smallest, at 10in (25cm), were too large for dolls' houses and were exhibited in suitably sized dolls' rooms.
Trademarks: *Huret* [above] *Exposition Universelle de 1885* [in a medal with Napoleon Bonaparte in the centre]; *Huret 34 Boulv Haussman Paris* [in three lines]; *Huret*; *Brevet d'Inv S.G.D.G.*

HURET
68 RUE DE LA BOETIE

Maison Huret Boulevard Montmartre, 22 Paris [in four lines]

Juillien (*jeune*)
40 rue de Beaubourg
Paris
1863 to early 20th century
Mostly large dolls but also many dolls' house dolls from c.1875 to 1880s, continuing with larger dolls. Became part of S.F.B.J. in 1904.
Trademarks: *J Jne* [in an oval]; *Jullien*; *Marque JJ*; *JJ*

JULLIEN

Les Arts du Papier
168 rue Vercingétorix
Paris
early 20th century
Dolls' rooms, dolls and associated items.
Trademarks: *La Mignonne* [beneath a rule]; *A.P. Paris* [in an oval]

Marchal et Buffard
Passage de l'Opéra
Paris
c.1850–c.1900
Dolls, furniture and associated items.

Marie et Bouquerel
Paris
1863–5
Exported dolls, rooms and furniture to several countries, including Britain.

Merle, M.A.
Paris
c.1850–c.1900
Room settings and dolls.

Merz, Emile
45 rue Said-Carnot
Beauvais
early 20th century
Dolls' furniture, room settings and dolls.

Trademarks: *E.M.B.* [in a circle]; *E.M.B.* [in a circle with jokers' masks]

Nadaud, A.
Paris
c.1870–c.1890
All types of dolls, including those for dolls' houses.
Trademarks: *Jouets Cotillion Nadaud 32 rue du Septembre* [in three lines in an oval]

Ousius, M.
Paris
from 1860s
Fretwork dolls' house furniture and room settings.

Péan Frères
Paris
c.1860–c.1890
Dolls' house rooms (including kitchens) settings and furniture.
In 1886 Péan Frères was a founder member of the Chambre Syndicate des Fabricants de Jouets Français.
Trademarks: *P.F.*; *Manufacture de Jouets en Metal Paris et Creil* [in a circle around] *P.F.*

Plichon, A.
Paris
c.1850–c.1900
Dolls' house dolls and accessories.

Prinoth et Cie
48 boulevard de Strasbourg
Paris
c.1875–c.1900
Dolls, dolls' house dolls and furniture.

Rossignol, Charles
Paris
1868–c.1900+
Dolls' house dolls, room settings and fittings.
Trademarks: *C.R.*

C.R.

Sussfeld et Cie, Société
21 rue de l'Echiquier
Paris
c.1860–c.1930
Distributors of all kinds of doll. The registered London office was c/o G.F. Redfern & Co., 15 South Street, Finsbury, London EC2.
Trademarks: *Clio Bébé Paris*; *Thalie Bébé Paris*

Unis France
Paris
1921–c.1930+
All sizes of dolls, including dolls' house dolls 2½in (6cm) high. The trademark "Unis France" was registered by the S.F.B.J. in 1921.
Trademark: *Unis France* [in an oval or in a diamond]

Verlingue, J.
Boulogne-sur-Mer and Montreuil-sous-Bois
1915–21
Dolls and dolls' house dolls.
Trademark: *Petite Française France J* [anchor] *V*

GERMANY

Bäselsöder, J.A.
Nuremberg
c.1858 to early 20th century
Tin furniture and kitchens.
Trademark: *JAB*

JAB

Baudenbacher, C.
Nuremberg

c.1835 to 20th century
Metal and wood toys and dolls; later production mainly in wood.
Trademark: *CBN* [with crossed wood-working tools]; registered in 1900

Becker, Gebrüder, & Glaser
Königsee, Thuringia
1897-1935 (possibly later)
Bisque and wood dolls, tea sets, dolls' house ornaments and fittings.
The company's London agent was W. Seelig (*q.v.*).

Bestelmeier, Georg Hieronimus
Nuremberg
1793–1854
Distributor of early wood dolls and bisque dolls' house dolls, kitchens, shops, tea sets, toys, early dolls' houses and appurtenances.

Bierhals, Carl
Nuremberg
c.1850–c.1930
Bisque and wood dolls' house dolls, kitchens, shops, tea sets, etc.
Trademark: *Leckermäulchen*

„Leckermäulchen"

Bing, Gebrüder
Nuremberg
c.1836–1933
Metal toys of all kinds, including kitchen stoves; one of Germany's most successful toy manufacturers; taken over by Bub in 1933. Dolls were added to the range in 1920.
Trademarks: varied over the years but usually incorporated the *G.B.N.* (until World War I) and *B.W.* (between 1917 and 1932)

BING-B WERKE

Blumhardt, H. & Co.
Stuttgart
19th century
Metal birdcages and similar ornamental items; also stoves, fireplaces, etc. displayed in good wood and metal room settings. Mentioned in reports of 1867 Paris Exhibition.

Brandt, Carl (junior)
Gössnitz, Saxony
c.1850–c.1930
Quality wood dolls' houses, kitchens, furniture, shops, room settings, dolls' house dolls, toys, building bricks, etc.
Trademark: a dolls' house; *C B jr* [intertwined]

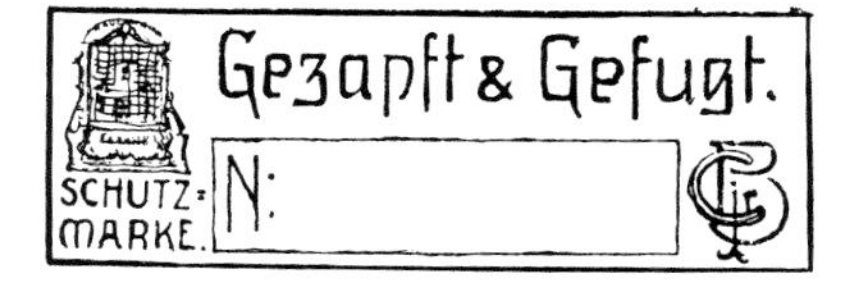

Brunner, Georg
Nuremberg
19th to early 20th century
Metal furniture, knick-knacks, goblets, tea services, etc.
Trademark: *Sanofix*

Sanofix

Buehrer, Frederick
Württemberg
19th century
Fully fitted kitchens and copper kitchen utensils.

Conta & Böhme
Pössneck, Thuringia
1790–1937
Bisque dolls, including dolls' house dolls and "frozen" Charlottes with clenched fists.
Trademark: a shield with a knight's arm holding a sword

Degenring, Theodor
Günthersfeld, Thuringia
c.1880–1939 (possibly later)
Miniature porcelain items, including knick-knacks and dolls' heads.
Trademark: crossed swords above *G* [in a circle]

Dieterich, C.F.
Ludwigsburg
*fl.*1850
Exhibited metal kitchens, furniture and bathroom items at the Great Exhibition in 1851.

Eck, Berthold
Unterneubronn, Thuringia
c.1876–82
Wooden dolls' furniture, room settings, shops, etc.
Trademark: stag's head in a shield

Eichner, G.L.
Nuremberg
19th century
Metal kitchens, kitchen utensils and bathroom fittings.

Geyer, Carl & Co.
Sonneberg
1882–1913
Bisque and wood dolls, dolls' house dolls, shops, room settings, doll toys, etc. Became Carl Geyer & Sohn in 1913.
Trademarks: four dolls in a decorated circle; *Bébé habille*; cornucopia [in a diamond]; *Liliput* [in a banner]

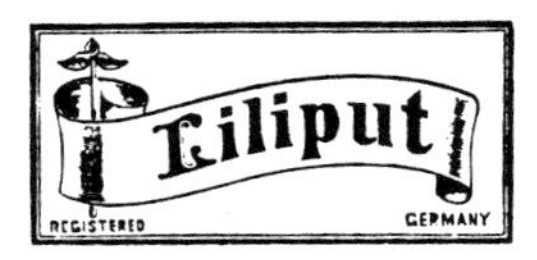

Goebel, William & F. & W.
Oeslau
1867–c.1930
Tiny and large bisque and porcelain dolls, ornaments and all manner of miniature items.
Trademarks: crown above *GW* [intertwined]; *GW* [intertwined] with size and mould numbers; new moon and triangle

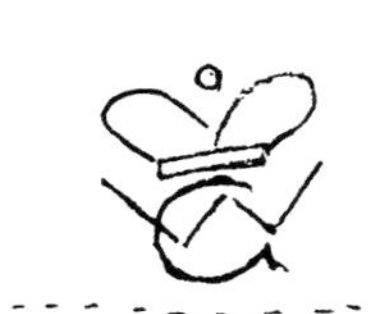

Hacker, Christian
Nuremberg
c.1875 to 20th century
Dolls' houses, furniture, shops, kitchens, etc. (See pages 41 and 137).
Trademark: *CH* [intertwined] within a crowned shield

Haller, J.
Vienna
1850
Wonderful display of toys, dolls and dolls' house furniture exhibited at the Great Exhibition in 1851.

Heber & Co.
Neustadt
1900–22
Bisque dolls, including dolls' house dolls, dolls' heads, knick-knacks and miniature items. Note the similarity of Heber's and Hacker's trademarks.
Trademark: *CH* [intertwined] in a crowned shield

Heinrichmaier & Wünsch
Rothenburg
c.1850–20th century
Dolls' houses, shops, kitchens, room settings, furniture and dolls.
Trademark: winged wheel above *H&W* [in an oval]

Hertwig & Co.
Katzhütte, Thuringia
1864–1939 (possibly later)
Most types of dolls, including dolls' house dolls and "snow babies".
Trademarks: *Hewika*; cat in outline of house with *H* in "attic" space, sometimes with *Hertwig* [in script]

Insam & Prinöth
Grödnertal (St Ulrich)
1810–1940 (possibly later)
Wooden dolls of all sizes from ½in (12mm) upwards, known as "Grödnertals"; also painted dolls' house furniture, etc.

Kestner, J.D.
Waltershausen, Thuringia
1805–c.1900
Bisque dolls of all sizes including dolls' house dolls. (See page 111.)
Trademarks: a variety of marks, including crown with *Kestner* on rim; crown with streamers; *J.D.K.*

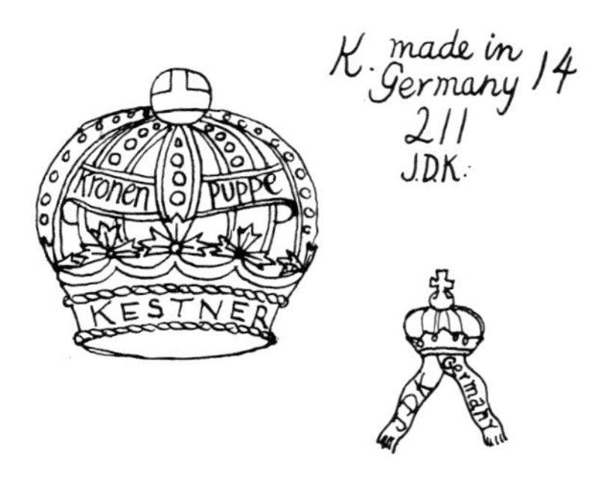

Kindler & Briel
Böblingen
1865 to 20th century

Wooden toys, furniture, small shops and room settings; also kitchens and stoves.
Trademarks: *Kibri*; Father Christmas carrying a sack of toys; *K & BB*

Kling, C.F. & Co.
Ohrdruf, Thuringia
1834–1940 (possibly later)
Good quality porcelain dolls of all sizes, some with moulded hair, including dolls' house dolls (See page 89.)
Trademark: *K* [in a bell, sometimes in a six-pointed star]

Kloster Veilsdorf
Veilsdorf on Werra, Thuringia
1765–c.1950
Bisque dolls including dolls' house dolls and "frozen" Charlottes; tea sets, ornaments, etc.
Trademark: *CV* [intertwined, sometimes in a circle]

Knosp & Backe
Stuttgart
19th century
Metal kitchens, rooms and accessories; wooden dolls' houses and room settings.

Kohnstam, M. & Co.
Furth
1876–c.1930
Distributed a wide range of toys, dolls' houses, dolls' house dolls, furniture and accessories. Branches in London, Milan and Brussels.
Trademark: *Moko*

Lange, Georg–Erben
*fl.*1850
Exhibited wooden dolls at the Great Exhibition in 1851.

Limbach AG.
Alsbach, Thuringia
1772–c.1939
Porcelain dolls, dolls' house dolls, ornaments and accessories.
Trademarks: several used, including crown and shield above *Limbach*; crown and clover leaf above *Limbach*; outline of standing doll above *Porzellanfabrik Limbach AG*

Cellulobrin,

Loewenthal & Co.
Hamburg
1836
Wooden dolls and papier mâché items; participated in the Great Exhibition in 1851.

Märklin, Gebrüder
Göppingen
1859–c.1930
The major manufacturer of tinplate toys of all kinds; took over Rock & Graner (*q.v.*).
Trademark: *G.M. Cie* [intertwined in a shield]

Matthes, E.W.
Berlin
1853–c.1930
Everything to do with dolls – houses, shops, dolls' house dolls, furniture, accessories, etc.
Trademark: *Friedel*

„Friedel"

Morgenroth & Co.
Gotha
1866–1919
Dolls' house dolls, shops, kitchens, ornaments, room settings and accessories.

Müller, Andreas
Sonneberg
1886–1930 (possibly later)
Dolls, dolls' house dolls and accessories.
Trademark: crown above crossed swords

Müller, B.A.
Dresden
Late 19th to early 20th century
Toys of all kinds, including bisque dolls' house dolls and wooden shops and furniture.
Trademark: toy soldier and teddy bear with a box full of toys

Pätzig, R. & Co.
Niederneuschönberg
early 19th century
Wooden toys, including furniture, shops and fittings.
Trademark: wreath surrounding

TOP:
Illustrated here is a selection of rare Rock & Graner dolls' house furniture made during the 1870s. At the top left is part of a set of drawing room furniture: a carver, one of the four side chairs and a sofa. All are tinplate and metal, simulated to resemble grained wood, with cabriole legs and upholstered in scarlet velvet edged with braid. The set also includes a small leather footstool (not illustrated). The sofa is 8¼in (21cm) long. The sofa and easy chair (one of four) to the top right are well upholstered over a tinplate framework. The printed cotton fabric is trimmed with woven braid, and the sofa is 7in (18cm) long.

At the left of the bottom is a tinplate dining table and a side chair. Both are in simulated grained wood finish and the chair has a simulated rattan cane seat. The table, which is raised on a central column with three cabriole legs, is 5in (15cm) tall. The wine table with its filigree gallery and the integral table and seats (in which are two tiny all-bisque dolls) are metal, simulated to resemble grained wood. The table with integral chairs is 3½in (9cm) wide. The lid of the washstand lifts to reveal the yellow-painted interior and a ceramic bowl, the cupboard beneath opening to disclose the ceramic chamberpot. The final item in this group is a chest of drawers with three long dummy doors, which form a hinged door. This opens to show the yellow-painted interior. Both this and the washstand are tinplate with a simulated wood grain finish.

Courtesy: Sotheby's, London

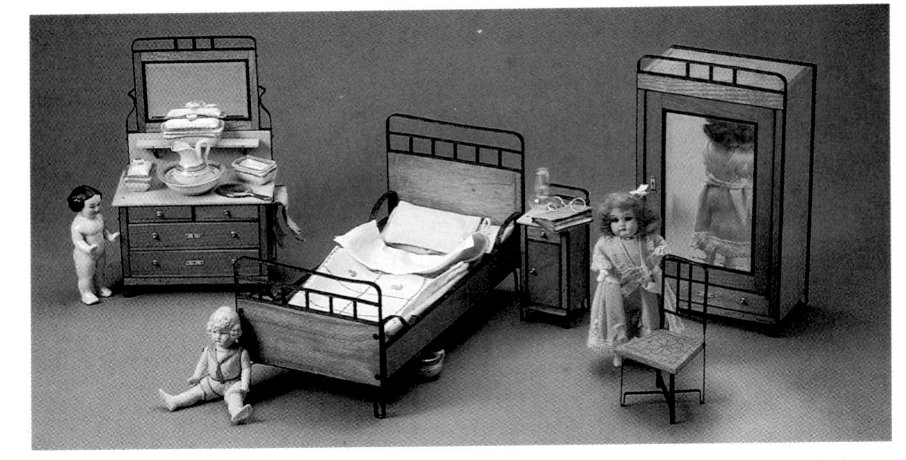

ABOVE:
This bedroom furniture – a wardrobe, bed, washstand-cum-dressing table, bedside cabinet and chair – are made of polished light oak (not simulated). The strong metal surrounds are painted black. Made between 1920 and 1925, the pieces are shown in the patent dated 10 July 1920, which was taken out by Bruno Ulbricht of Nuremberg. The wardrobe is 11in (28cm) high. The three dolls are German and date from 1875–1925. The smallest, 5in (14cm) tall, is the *Badekinder* (bathing doll), the seated doll is 5¼in (13cm) tall, and the standing doll, made by Heubach of Koppelsdorf, is 7¾in (19cm) high. See also pages 71 and 123.

Author's collection; photograph: Frank Newboulc

shield over crossed swords, *RPC* [monogram] in shield

Pensky, Alfred
Steinbach
1919–38 (possibly later)
Dolls' house dolls, bathing dolls, "half" dolls and "piano" dolls.

Reusz, Erna
Berlin
early 20th century
Dolls' house furniture, room settings and dolls' accessories.
Trademark: octagonal device containing *ER*

Rock & Graner
Biberach an der Riss
1837
Exceptionally fine quality tin furniture. (See pages 83 and 122.) Also made wooden room settings, dolls, dolls' house and tinplate kitchen equipment. There were several successors to this company, which was thereafter known as Rock & Graner Nachf. (*Nachfolger* = successors). Finally taken over by Märklin (*q.v.*).
Trademark: included initials *R. & G.N.*

Salner, Joseph
Zwickau
c.1890–c.1930 (possibly later)
Wooden furniture, rooms, shops and dolls; possibly some dolls' houses.
Trademarks: *Modestes*; *Dreistern* [in a circle around three stars]

Samulon, J., Nachf.
Dresden
1892–c.1930
Good quality dolls' house dolls and "family" groups.

Schindhelm & Knauer
Sonneberg
1919–1930s
Dolls' houses and rooms, dolls' house dolls, toys, etc.

Schmidt, Bruno
Waltershausen
1898–c.1930 (possibly later)
Bisque dolls, including dolls' house dolls. (See page 90.)
Trademarks: *B.S.W.* [in a heart]; *Herz*; a heart

Schneegass
Waltershausen bei Gotha
(Eventually became Gebrüder Schneegass & Söhne)
Late 1830s to 20th century
Exceptionally fine quality dolls' house furniture in simulated rosewood and Regency and art nouveau styles; also dolls' house dolls. This well-respected company continued to produce "Waltershausen" furniture into the 20th century, and many examples of the style are illustrated in this book. Other manufacturers in the area also made good quality dolls' house furniture and furnishings.

Schubert, Hermann
Berlin
1885–c.1930 (possibly later)
Dolls' houses, rooms, shops, kitchens and dolls.

Schulz, Joseph
Meiningen
19th century
Ivory toys and furniture.

Schweitzer, Babette und Familie
Diessen am Ammeisee
end of 18th century to present
Silvered soft metal, filigree furniture, also painted miniature perambulators, vases, candlesticks, bird cages, etc.

Söhlke, M.G.
Berlin
*fl.*1850
Exhibited gilded tin dinner and tea services, stylish dolls' house furniture of good quality, stoves, fireplaces, etc. at 1851 Great Exhibition and other European exhibitions.

Ulbricht, Bruno
Nuremberg
late 19th to first half 20th century
Patented dolls' house furniture suitable for large dolls' houses, and also rooms, toys, etc. (See pages 36, 71 and 122.)
Trademark: *Ulgo*

77
BIRD'S
CUSTARD POWDER
GENERAL

Left-hand house

Date **1920s**
Maker **Unknown German**
Height **23in (58cm) (including base)**
Width **20in (51cm)**
Depth **15in (38cm)**

This German blue-roof house has lithographed "stone" and "red brick" paper over the exterior. The front opens to reveal two rooms, both with their original wall and floor coverings.

Centre house

Date **1920s**
Maker **Unknown German**
Height **18½in (47cm) (including base)**
Width **13in (33cm)**
Depth **8½in (21cm)**

This delightful blue-roof house has a balcony and front porch to add interest to the façade. It is front opening and has two rooms; note especially the detail added to simulate a basement.

Right-hand house

Date **Late 19th century**
Maker **Possibly Silber & Fleming**
Height **22in (56cm)**
Width **11in (28cm)**
Depth **9in (22cm)**

The name Kosy Kot is painted on the front of this house, and the name plate on the door bears the name *G.* [or perhaps *C.*] *Fisher*. There is a letter box beneath the name plate. The exterior is partly painted and partly papered to simulate red brick, although this has now faded. The lower part of the front would originally have been papered to simulate stone. There is a large advertisement for Pears soap on the side. The interior of this house is illustrated on page 9.

The tiny "red brick" house in the foreground, which is only 8in (20cm) high, 5½in (14.5cm) wide and 3½in (9cm) deep, is an English house, by either Lines or Silber & Fleming. The hansom cab (front left), Schuco drummer and miniature train set are from the collection of the Abbey House Museum, but other items, including the tiny model of a Bliss house, the two small German boy dolls and the doll passengers on the bus, are from the author's collection.

Courtesy: Abbey House Museum, Kirkstall, Leeds; photograph: Frank Newbould

Winkler, Ernst
Sonneberg
1903–27
Dolls, dolls' house dolls and many novelties.

BRITAIN

Ashcroft, Messrs
Liverpool
19th to 20th century
Although chiefly manufacturers of billiard tables, also made dolls' houses.

Avery, W. & Sons
Redditch
19th to 20th century
Strong, painted metal but good quality dolls' house furniture, some gilded.
Trademark: *W. Avery Redditch*

Baetons, Pauline
18 Oxendon Street
London EC
*fl.*1850
Exhibited miniature boxed playing cards, ½in (12mm) long, at the Great Exhibition in 1851.

Barton, A. & Co. (Toys) Ltd
Morden, Surrey
later, Easthill, London
finally, New Addington, Surrey
c.1940 to present
Dolls' house furniture; supplied "Grecon" dolls (*see* Cohn Margarete).

Bouchet, A.
London
*fl.*1850
Described as exhibiting dolls and shops at the Great Exhibition in 1851.

Britain, William
London
mid-19th century to present
Although later production was devoted to lead figures and military items, did supply some dolls' house items.
Trademark: *W. Britain*

Bull, Thomas Henry
Newington Causeway, Surrey
19th century
Pewter toys, kitchens and tea sets.
Trademark: table set with "pewter" cups and coffee pot

Child, Coles
London Bridge
c.1750 to late 18th century
Pewter and tin toys and other items.

Cohn, Margarete (Grete)
40 Streatleigh Court
London SW10
20th century
Dolls with wire armatures, covered in wool, with composition heads and feet, details embroidered in wool. Her dolls are to be found in many dolls' houses of the 1930s to 1960s from both Britain and Germany. Her trademark was registered in Berlin in 1920 and in London in May 1940 (no.611 679).
Trademark: *Grecon*

Cole & Co.
6 Falcon Square
London
19th century

Brass, copper, tin and other metal dolls' house items, kitchens, etc.
Trademark: a coal scuttle

Cooper, Frederick & Sons
Holborn
London
end 19th to early 20th century
Dolls' houses and dolls.

Cremer & Sons
London
19th century
Toys of all descriptions. An article in 1865 in *English Woman's Domestic Magazine* referred to Cremer as the "Merlin" of the toy trade, and in 1873 Cremer described his shop as being stocked with "a goodly selection of toys", among which were dolls' houses, rooms, furniture and dolls.
Trademark: *Cremer Dolls Toys Games 210 Regent Street* [in an oval]

Dear, John Cox
Bishopsgate-Without
*fl.*1850
Exhibited dolls, toys and furniture at the Great Exhibition in 1851.

Dol-Toi Products (Stamford) Ltd
Lincolnshire
1940s on
Dolls' house furniture and dolls.

Edlin's Rational Repository of Amusement Toy Shop
London
1810–c.1840
A great variety of toys.

Elgin, Eric
Enfield
1919–26 (See pages 66, 67 and 81.)
Dolls' house furniture.

Date **1875–1900**
Maker **G. & J. Lines**
Height **30in (76cm) (excluding base)**
Width **38in (96cm)**

A strongly constructed dolls' house produced by G. & J. Lines during the last quarter of the 19th century, but of a type continuing to be made (in common with most Lines houses) into the 20th century. This particular style was "modernized" at the beginning of this century by two additions – a flat-topped, pitched roof with two dormer windows and a garage, which was attached to the base at one side. The new style became No. 25 in Lines' catalogue (see page 135), but the house illustrated here is the forerunner of No. 25. It has a flat roof and a centrally placed door, flanked by columns and surmounted by a small balcony. There are double bay windows on the ground floor, and the front opens to reveal four rooms. I have seen many adapted versions of the house seen here, some adaptations made by professional carpenters, others by handymen, but many given the most impressive front elevations. Sharp-eyed collectors should look out for such extensions of the basic four-room model. The house illustrated here was repainted at some time; its lower half would originally have been papered in "red brick".

Courtesy: Sotheby's, Chester

Four items of period dolls' house furniture manufactured by Lines Brothers Ltd of London. Three are shown on top of their original boxes, which bear the original prices: the chest of drawers cost 6s 6d, the settee, 2s 6d and the oak chest, 1s 9d. The knole-style red velvet and red and gold braid-covered settee, which is resting on top of the oak chest, is also one of the numerous items produced by this company during the first half of the 20th century. It is difficult to date the furniture exactly as the same styles were produced between the two World Wars and even after 1945.

Author's collection; photograph: Fabian

Emell Toy Mfg. Co.
Islington, London
1915–22
Dolls' furniture, dolls' house dolls, a dolls' house named "palace", toys, shops and folding dolls' houses.
Trademark: *ML*

Evans, Joseph & Sons
London
c.1850–c.1900
Dolls, dolls' house dolls, boxed furniture and tea services.
Trademark: *Joseph Evans & Sons* [and address in an oval]

Foster, Blackett & Wilson Ltd
London
20th century
Dolls' furniture, shops and small dolls.

Goodman, L. & A.L.
London N16
c.1900–50
Importer of toys, including dolls' house furniture, dolls and other miniature items.

Gray & Nicholls Ltd
Liverpool
c.1900–50
Dolls of all sizes, furniture and miniature items.

Higgs, William
London
*fl.*1733
Maker and turner of wooden dolls; may also have made miniature chairs.

Hyatt, Joseph
London
end 19th to early 20th century
Imported dolls, furniture and other miniature items.

Iliff, W.J.
Newington, London
*fl.*1850
Exhibited small gutta percha dolls and other miniature items at the Great Exhibition in 1851.

Jones, Robert
Great Mitchell Street
London
1835–c.1850
Wooden dolls and miniature items. In 1850s Robert was succeeded by Henry Jones.

Lascelles, Edward
Wavertree
*fl.*1850
Exhibited dolls' houses at the Great Exhibition in 1851.

Lascelles, J.W.
Liverpool
*fl.*1850
Exhibited a model of a mansion at the Great Exhibition in 1851.

Leuchers, William
London EC
established before c.1850
Toys, shops, furniture, dolls and other miniature items.

Lines, G. & J. Ltd
457 Caledonian Road
London
1858–1919
Dolls' houses and furniture.
Trademark: *G & J Ltd London* [in a circle around a thistle]

Lines Brothers
London
1919–c.1970
Dolls' houses and furniture.
Trademarks: *Tri-ang* (registered 1927); *Minic* (registered 1930s)

Lucas, Henry
London
*fl.*1850
Exhibited miniature items at the Great Exhibition in 1851.

Mitchell, J.
Stonehaven, Scotland
*fl.*1850
Exhibited small items for dolls' houses at the Great Exhibition in 1851.

Morrell's
Burlington Arcade
London
c.1820–1950 (moved to Oxford Street in 1892)
Finest quality items for dolls' houses in ivory, porcelain, silver and wood.

Poole, John & L.
Twister's Alley
London
c.1840–c.1860
Wooden dolls and furniture; may also have made dolls' room settings.

Powell, J.
Trentham
Newcastle-under-Lyme
*fl.*1850
Exhibited a model of Shakespeare's

The Tri-ang label from the left-hand house illustrated on page 136. Look for the label on houses produced by Lines Brothers over a period of some thirty years. It is sometimes at the top of the gable end (as here), sometimes near the base. The label may have been obscured if the house has been repainted, and, if you suspect this, it is worth scratching the paint carefully to see if you can reveal the label.
Author's collection; photograph: Fabian

birthplace made of oak and plaster at the Great Exhibition in 1851.

Remill, Misses
London
*fl.*1850
Exhibited 20 miniature items of bone and tragacanth at the Great Exhibition in 1851.

Robinson, Francis K.
Whitby
*fl.*1850
Exhibited miniature items, some of jet, at the Great Exhibition in 1851.

Russell, Robert
Tunbridge Wells
*fl.*1850
Exhibited miniature items at the Great Exhibition in 1851.

The Rustless & General Iron Co.
97 Cannon Street
London
mid-19th to 20th century
Tin and metal kitchens and kitchen utensils.
Trademark: *Anti-Corrodo Tr & Gl Co. WB*

ANTI-CORRODO
TR & GI
CO
WB

Seelig, William
London
early 20th century on
Dolls' houses, dolls' house furniture, dolls, etc.
Trademark: *Everrest*

Silber & Fleming
London
c.1850–1900
Dolls' houses in a variety of sizes and styles and at a wide range of prices.

Sillett, John
Kelsale
Saxmundham
*fl.*1850
Exhibited a single-storey dolls' house in 1851.

Spurin, E.C.
37 New Bond Street
London
before c.1850 to end 19th century
A wide variety of toys, including dolls, dolls' houses, room settings, furniture, etc. The agent for Fleischmann when the "model of Gulliver in Lilliput" was staged at the Great Exhibition in 1851.

Tattersall, J., Ltd
Southport
first half 20th century
Dolls, dolls' houses, furniture and miniature items.

Trebeck, Thomas Frederick
London
*fl.*1850
Exhibited dolls, dolls' houses and furniture at the Great Exhibition in 1851.

Turnbull, C.E. & Co.
London EC
c.1875–1925 on
Dolls' houses, dolls and furniture.

Wicks, William
Islington, London
*fl.*1852
Wooden dolls and furniture.

AMERICA

Althof, Bergmann & Co.
New York City
1867 to 20th century
Made finely crafted tin dolls' house furniture and kitchen utensils, and imported dolls' house knick-knacks and so on.
Trademark: *A.B.C.* (registered in 1881)

Arcade Manufacturing Co.
Freeport, Illinois
founded c.1850
Made metal furniture for dolls' houses.

Bliss, Rufus, Manufacturing Co.
Pawtucket, Rhode Island
founded 1832
One of the most prolific of US manufacturers; output included lithographed houses, folding houses (*see* page 48), furniture and so on. Supplied many stores throughout the US.

Butler Brothers
New York City
founded 1877
Founded in Sonneberg, Germany, as both a doll manufacturer and distributor. During latter part of 19th century one of the major distributors in the US, with a head office in New York City. Products of Bliss and Converse (*qq.v.*) among others in its catalogues.

Cissna, W.A. & Co.
Chicago
Major store/wholesaler stocking products of Bliss, Converse (*qq.v.*) among others.

Converse, M.E. & Son
Winchendon, Massachusetts
founded 1878
Wooden toy manufacturers, especially famous for dolls' houses and other structures; a similar product to Bliss and stocked by many of the same stores and wholesalers.

Dowst Manufacturing Co.
Chicago
founded c.1875
Made metal dolls' house furniture and dolls' "mansions", which were made from a type of heavy pressed board. Also made room settings. In the 1920s began to produce items under the tradename *Tootsietoy*. Dowst products are sometimes confused with those of Tynietoy (*q.v.*).
Trademark: *Tootsietoy* (registered early 1920s)

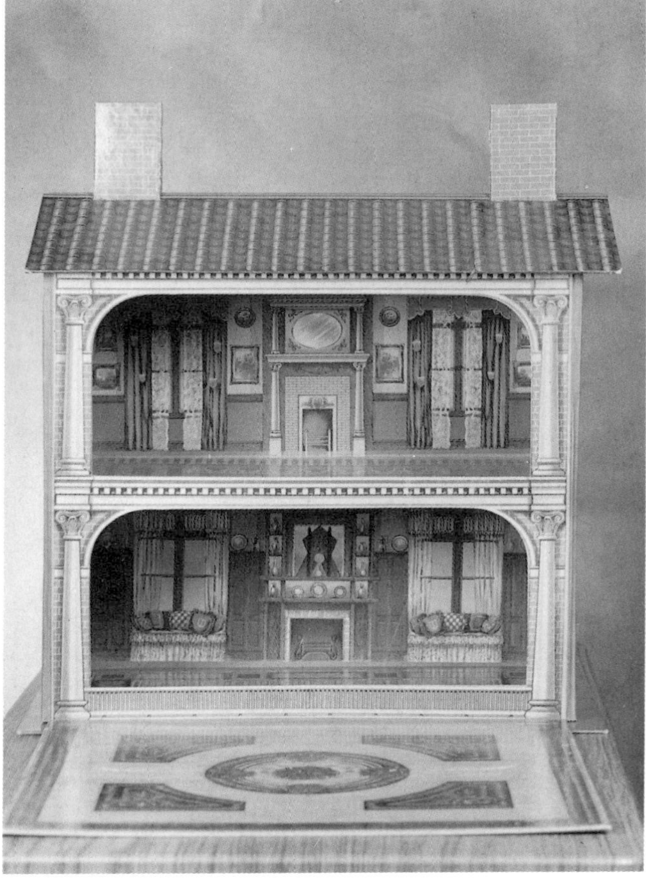

Ellis, Britton & Eaton
Vermont Novelty Works
Springfield, Vermont
founded 1860s
Furniture, dolls' household utensils and many finely designed articles, including an office set in iron, cleverly painted to simulate brown "graining" on a fawny-yellow base. There is a set in the Wenham Historical Museum, Massachusetts, set out as a lawyer's office.

Francis, Field & Francis
Philadelphia
founded c.1838
Made every conceivable type of household furniture for miniature dwellings, metal dolls' houses and a range of simple tin toys. Some of the dolls' house furniture is comparable with that of Rock & Graner (*q.v.*) although it is slightly heavier.
Trademark: company name

Hull & Stafford
Clinton, Connecticut
founded c.1860
(Originally Hull & Wright.) Made a variety of mechanical toys of light-weight construction.

McLoughlin
founded 1855
Especially famous for lithographed folding houses. (See above)

Schoenhut, A. & Co.
Philadelphia
founded 1872
An emigré from Germany, Albert Schoenhut founded a company that became particularly famous for wooden toys – toy pianos, circuses and so on. During World War I the company began to manufacture dolls' houses (possibly because shipments from Europe were interrupted); these houses were sturdy rather than elegant; most were bungalows and modern in design.

Stevens & Brown Manufacturing Co.
Connecticut
founded 1868
Formed by the merger of George W. Brown & Co. and J. & E. Stevens, the company made cast iron toys and tinplate dolls' house furniture; room settings with "plush" uphol-

stered furniture and sturdy iron furniture, similar to that produced by Francis, Field & Francis (*q.v.*).

Stirn & Lyon
New York
founded c.1880
Patented folding dolls' houses and other buildings, advertised as "combination mansions". Took out a patent on 11 April 1882.

Tynietoy
Providence, Rhode Island
founded 1920s
Quality furniture and dolls' houses. This company's products are sometimes confused with those of Dowst Manufacturing Co. (*q.v.*), whose tradename was *Tootsietoy*.

Date **c.1911**
Maker **McLoughlin Brothers**
Height **19½in (50cm)**
Width **18in (46cm)**
Depth **9½in (24cm)**

The McLoughlin garden house is one of the company's highly collectable lithographed folding houses. The colours of this house are as bright and fresh as when it was first made. McLoughlin Brothers of New York became established in the mid-19th century and listed folding dolls' houses in a catalogue as early as 1875. The instructions for this model, the "New Folding Doll House", are inside the top cover of the box. "All parts of the house are hinged together except the roof and chimneys. Besides these it consists of a back wall, two side walls, two floors, an inner front wall partly open, and an outer front wall which can be let down to form a courtyard. To set the house up, take it as it comes folded and hold it on edge so that it stands properly upright. The lower floor is folded up against the back wall; let this floor drop and then reverse it so that it is brought under the house. Then let the outside front drop, and pull it forward. This will draw the side walls (which have a hinge in the middle) into a straight position. Then the upper floor can be pushed down in place, and when the roof and chimney are put on, the house will stand complete."

Courtesy: Mrs Elizabeth M. Donoghue, Everett, Massachusetts

PRICE GUIDE

DOLLS' HOUSES

There are so many imponderables where dolls' houses are concerned that it is impossible to do more than give the most general guidance. At one end of the scale are unique baby houses of historical interest and importance, at the other card and paper houses given away for advertising purposes in the 19th and early 20th centuries. There are, on the one hand, some good, late, commercially made houses and shops and, on the other, some late, home-made, crudely converted boxes of little value. Sometimes the contents, if of good quality, may be more valuable than the house. As always, condition affects the price.

PRICE RANGE	PERIOD	TYPES OF HOUSE
A £25,000–£35,000	17th to early 18th century	Exceptional cabinet or baby houses, with provenance and with original contents
B £20,000–£30,000	17th to early 18th century	As above, but lacking contents or having only a very few minor items
C £15,000–£20,000	c.1730–80	Baby houses in reasonably good condition with some original and some later contents
D £8,000–£15,000	c.1730–80	As above, but unfurnished
E £6,000–£10,000	c.1730–1830	Cabinet or baby houses, according to condition and the quantity and quality of contents, merging into houses that have clearly been played with and might be called dolls' houses
F £5,000–£10,000	Late 18th century to c.1830	Baby houses in reasonably good condition, with some contents
G £5,000–£10,000	Late 18th century to end of 19th century	Dolls' houses in good condition, preferably with provenance, and with original fixtures, furniture, and dolls of the same period, façade "all original". (If with original stand, an exceptional item could be lifted into the B range)
H £500–£3,000	Late 18th century to end of 19th century	Dolls' houses of merit and without fault, but unfurnished, and houses in only fair condition but with some early furniture and/or good fittings, such as chandeliers
I £300–£2,500	Mid to end 19th century to 1950s	Commercially made dolls' houses, with a wide range of façades – lithographed or painted – with contents of varying quality (some furniture, furnishings, or dolls may be from an earlier period or may be hand made), joiner-made dolls' houses, and folding dolls' houses

J £25– £300	Mid to end 19th century to 1950s	As above, but houses in poor (or worse) condition, perhaps with structural faults or deficiencies, with contents of little or no value or no contents at all

Date End 19th century
Maker Unknown American
Height 45in (114cm)*
Width 50in (127cm)*
Depth 25in (64cm)* (at deepest part)
(*Dimensions exclude base)

A large, solidly built residence of sober character but with intriguing features on the exterior – raised wooden strips in a pattern that is neither "quoining" in the true sense of the word nor "rustication". One could say that the decorative wooden strips are a mixture of outlining and ornamentation. The exterior is painted in two shades of green, and the doors and window frames are painted brown. There are 15 single and three double windows, and the front of the house is hinged to open in three sections to reveal six rooms. The wooden floors are painted in a variety of designs. The owner believes that the house was originally bought from F.A.O. Schwartz but has no documentary proof. The curtains are original. American dolls' house collectors call houses such as this "mystery" houses, and other examples may be seen in the Margaret Woodberry Strong Museum. *Private collection*

DOLLS' HOUSE DOLLS

PRICE RANGE	PERIOD	TYPES OF DOLL
£400 to £900 or more	Late 18th to early 19th century	Poured or carved wax dolls, in original attire; wooden dolls; tragacanth dolls (clothes and condition affect value)
£100 to £300 or more	Late 18th to early 19th century	Wooden dolls – early Grödnertal (pointed chins, tiny waists)
£70 to £400 or more	Late 18th to early 19th century	Papier mâché dolls – kid bodies command the highest prices
£100 to £350 or more	1830–1850	China-head dolls on fully articulated wooden bodies, in original dress; wax dolls; early bisque dolls

£10 to £200	1850–1900	China-head, bisque, stone bisque, and wax dolls. Dutch-type dolls (rough-and-ready descendants of the elegant Grödnertals). Much depends on general quality of doll and clothing, so there is a very wide range of prices
£200 to £500 or more	1850–1900	Elegantly dressed mignonettes; china- or bisque-head dolls with military head-dress (prices depend on costume)

DOLLS' HOUSE FURNITURE

PRICE RANGE	PERIOD	TYPES OF FURNITURE
Very expensive	18th century	Choice, genuine, extremely rare items
£300 to £1,000 or more	1800–1830	Genuine, rare items
£100–£1,000 or more	1830 to early 20th century	Commercially manufactured furniture – Beidermeier, "Duncan Phyfe", and Waltershausen (the generic term used throughout this book). Production continued in this style by various German manufacturers, in simulated rosewood and other woods, with gilt ornamentation, throughout the period (see illustration). The highest-quality items were made by, among others, the Schneegas Co. (Waltershausen), and these command the highest prices. Tin, iron and alloy furniture such as that made by Rock & Graner (see pages 83 and 122). Gilt, ormolu furniture – the so-called Tiffany – also commands high prices. Bone or ivory furniture of dolls' house size. French faux-bambou furniture
£20–£400	Still in production or otherwise	Diessen; various German, French and American furniture manufacturers (prices vary widely; complete sets most expensive)
£20–£300	Early 20th century to c.1950s	Elgin furniture, marked (see page 97). The higher price would be for a complete, marked *set* in original condition. Unknown makes of medium quality furniture.
£5–£125 or more	Early 19th century to c.1950s	Lines Triang furniture in various styles (see page 127). Again the higher price level would be for a complete set, more if in original box Barton furniture (post-World War II and still in production)

ORNAMENTAL ITEMS

PRICE RANGE	PERIOD	TYPES OF ORNAMENT
£20–£600 or more	All periods	Birdcages, clocks, mirrors, pictures and artefacts; prices depend on period, quality, condition and rarity

GLOSSARY

Alcove: recess opening out of a room or into a wall.

Arch: curved structure, often ornamental, spanning an opening and supporting the wall above.

Architrave: the lowest division of an entablature surmounting a column; the moulding around doors, fireplaces and windows.

Aumbry: originally a cupboard in a church to hold an alms box, always with a door; now, generally, a cupboard in a wall.

Balachin: structure within a building in the form of a canopy.

Baluster: carved column or pillar supporting a handrail.

Balustrade: a series of balusters topped by a rail.

Bargeboard: board, often ornamental, concealing the ends of roof timbers.

Bay: a main division of a façade or interior wall.

Biedermeier: of the period 1810–45, the word is applied to dolls' house furniture and dolls with papier mâché shoulder-heads, varnished hair and painted features.

Bisque (sometimes known also as biscuit)**:** a ceramic material that can be poured into a mould or pressed into shape before being fired at high temperature; it is usually painted before being fired for a second time at a lower temperature and has an unglazed, matt surface.

Bolection moulding: a moulding projecting from an

Date Early 20th century
Maker G. & J. Lines
Height 40in (102cm)
Width 42in (107cm)

This well-made house dates from before World War I. It is No. 25 in Lines' catalogue, and it is worth comparing this house with the one illustrated on page 127. G. & J. Lines described this house as "a country residence with garage attached and beautiful garden laid out with flowers". In fact, the base tray pulls forward to form the balustraded formal garden with the steps leading up to the front door. The house is finely decorated throughout and has a staircase. When it was made, the house cost 100 shillings.

Courtesy: Christie's, South Kensington; photograph: A.C. Cooper

Left-hand house

Date **1930s**
Maker **Lines Brothers (Tri-ang)**
Height **16½in (42cm) (including base)**
Width **13¼in (34cm)**
Depth **10½in (27cm) (at longer side)**

No. 60 in Lines' catalogue, this house has the Tri-ang mark on the gable at the back of the house (see page 128). The catalogue describes it as follows: "two large rooms, front opening, nicely flowered, side porch and seat, metal windows."

Right-hand house

Date **c.1910**
Maker **Unknown German**
Height **21½in (55cm)**
Width **17½in (44cm)**
Depth **9½in (24cm)**

This red-roof German house offers an interesting contrast with the Lines house. It has three painted windows at each side and two bay windows on the ground floor at the front. On the first floor, set back from the balcony, two windows flank a French window. All the windows have painted shutters. The roof, with two dormer windows, is hinged and may be lifted to disclose the three-storey interior. Houses such as this were made in Germany for the English, French and American markets. It has been suggested that this house was made by Christian Hacker.

Author's collection; photograph: Frank Newbould

otherwise flat surface, a projecting moulding surrounding a panel.

Breveté (French): patented; abbreviated to *Bte*.

Bust head: the phrase used in the United States instead of shoulder head.

Chamfer: a bevelled edge.

China-head: a term to describe a doll with a head of glazed porcelain, a ceramic material that was widely used for dolls' heads before being largely superseded by bisque.

Composition: the name given to a variety of materials (including papier mâché) used to make dolls' heads, bodies and limbs and other toys.

Contemporary: the word used to describe items that are of the same period but not necessarily original to the subject – e.g., clothes of the same period but not original to a doll.

Corbel: a bracket of stone or timber, often decorative, projecting from a wall and acting as a support or bracket.

Cornice: a decorative feature high on a wall; the projecting upper portion of an entablature.

Curtail: the lowest step in a flight of stairs, ending in an outward-facing curve or scroll or elaborate ornamentation.

Dentil: a rectangular projection in a classical style cornice; a decorative feature on furniture.

Deposé (French) or **Deponiert** (German): the mark used to indicate that a manufacturer's trademark or patent has been registered; often abbreviated to *Dep* or *D.E.P.*

Dovetail: a fan-shaped tenon fitting into a mortice to form a joint.

Dowel: a cylindrical piece of wood used in joinery to effect joins.

D.R.G.M. (Deutsches Reichsgebrauchmuster): the initials used, from 1909, to indicate that a design or patent was registered in Germany.

Entablature: in classical architecture, the horizontal post – comprising the architrave, frieze and cornice – resting on the columns.

Escutcheon: protective, often decorative, metal plate around a keyhole.

Fanlight: a semi-circular window, with radiating glazing bars, over a door.

Feldspar (Felspar): a white or flesh-coloured mineral used in the production of non-translucent paste in the ceramic industry.

Fillet: a piece of wood used to fill in a space between one section and another.

Folding dolls' house: the term used to describe any dolls' house (or room) that may be taken apart and re-assembled.

French slope: the term used to describe the steeply cut-away crown of a dolls' head, usually fitted with a cork pate.

Frieze: a horizontal decoration at upper level of a wall.

"Frozen Charlotte" (or "Frozen Charlie"): a name given in the United States to a doll made in one piece, with head, body, arms and legs moulded together, they are also known as bathing dolls, or bathing babies (*Badekinder*), pillar dolls and solid chinas; they are made in bisque or glazed china.

Gambrel: a roof with two slopes on each of two sides, the lower steeper than the upper.

Gibb door: a disguised door, often made to look like bookshelves.

Grödnertal: the word used to describe wooden, slim-bodied dolls, with joints, which originated in the Groden valley in Germany in the early 18th century but which were further developed by German wood-carvers who moved to the Tyrol and Italy. By the end of the 19th century the type had degenerated into the form known as "Dutch" dolls – not because they came from the Netherlands but from a corruption of "Deutsch". The smaller sizes suitable for dolls' houses were also called "penny woodens".

Hipped roof: a roof whose ends and sides have the same slope or pitch.

Lintel: a horizontal support bridging the opening over a door, fireplace or window.

Mansard roof: a roof with two pitches, the lower steeper than the upper, designed to give space for a garret in the upper part of a house.

Milliners' models: a term used (without solid foundation but now firmly established) to describe dolls with papier mâché shoulder-heads, varnished hair (often elaborately styled) and painted features.

Mitre: a joint between two pieces of wood, made by cutting each to the same angle.

Mortice: a groove cut into stone or timber to receive a tenon and join two members together.

Newel post: a vertical post, often decorative, supporting the handrail at the foot of a staircase or on a landing, or about which a circular or spiral staircase winds.

Niche: a rounded, decorative recess in a wall, often flanking a fireplace.

Nosing: the rounded edge to a step or flat surface.

Papier mâché: a paper pulp (sometimes with admixtures) combined with a whitening agent and suitable glue, that was used for the manufacture of dolls' heads and bodies and other items in the early years of the 19th century. Towards the end of the 19th century a new type of papier mâché was developed: this could be poured into a mould rather than having to be pressure moulded, and it was stronger and more durable than the earlier substance.

Parian: true Parian was never used for doll making, but imitation Parian was sometimes used for dolls' heads; it was a pure white bisque on which the features were usually painted, although the eyes were sometimes glass.

Pediment: a triangular functional and ornamental finish to the tops of doors or of a façade.

Peg wooden doll: late 19th and early 20th century Grödnertal-type doll of debased construction and indifferently painted features.

Pitch: the slope of a roof.

Porcelain: used as a synonym for china when describing glazed dolls' heads and limbs, tea-services and so on.

Portico: a row of columns forming a colonnade in the front of a building or two columns forming a porch.

Poured wax: the term used to describe dolls manufactured by wax being poured directly into a mould, any residue being poured off to leave only a shell.

Presepio: a Neapolitan Christmas crib or crêche scene.

Provenance: documentary evidence of the history of a dolls' house, doll, furniture or other item, passed on by the original owner's friends or family.

Queen Anne: the term used to describe wooden dolls, furniture and other items, dating from the end of the 17th to early 18th century: it is often wrongly used to describe items manufactured as late as 1775.

Reproduction: the word used to describe a copy made in the likeness of another object, especially furniture, and also used to describe a doll made from a mould taken from the original doll or doll's head.

Riser: the upright portion of a step or stair.

Rustication: on dolls' houses, wood carved or affixed to the exterior walls to resemble stone.

S.F.B.J. (Société Française de Fabrication des Bébés et Jouets): a syndicate formed in 1899 by the best known French doll manufacturers to counter the threat from German doll manufacturers.

S.G.D.G. (Sans Garantie du Gouvernement): initials found on items of French manufacture signifying that the patent or trademark has not yet been registered – literally, that it had not yet been guaranteed by the French government.

String: the inclined timber by which the treads and risers of a staircase are supported.

String course: a moulding or projecting course, usually in stone, sometimes ornamental, running horizontally across the exterior of a building.

Tenon: in joinery the projecting piece of wood that fits into the mortice.

Tragacanth: a white or reddish gum derived from shrubs of the *Astragalus* genus found in western Asia.

Tympanum: the triangular surface bounded by the horizontal and sloping cornices of a pediment; also the space between the lintel and the arch of a medieval doorway.

Unis France (Union Nationale Inter-syndicate): syndicated with S.F.B.J.

Venetian window: a window consisting of a central, arched light flanked by two smaller lights.

Wainscot: decorative wooden panelling on the lower part of an interior wall or, simply, any decorative lining of an interior wall.

Winder: a step that is narrower on the inside than the outside, forming part of a spiral or winding staircase.

BIBLIOGRAPHY AND RECOMMENDED READING

Ackerman, Evelyn, *Victorian Architectural Splendor in a 19th Century Toy Catalogue*, Era Industries Inc., Culver City, California, 1980

Addy, Sydney O., *The Evolution of the English House*, George Allen & Unwin, 1898

d'Allemagne, Henri René, *Histoire des Jouets*, Hachette, Paris, 1903

—, *Musée Retrospectif de la Classe 100 Jouets l'Exposition Universelle International*, Paris, 1900

—, *Les Jouets à la World's Fair en 1904 à St Louis*, Paris, 1908

Anka, Georgine and Gauder, Ursula, *Die deutsche Puppenindustrie 1815–1940*, Puppen und Spielzug, Stuttgart, 1978

Barton, Lucy, *Historic Costume for the Stage*, Walter R. Baker & Co., Boston, Massachusetts, 1935

Beeton, Isabella, "Children's Fancy Work" in *Beeton's Book of Needlework*, Ward, Lock & Co., London c.1860

Braun, Hugh, *The Story of the English House*, B.T. Batsford Ltd, London, 1940

Calmettes, P., *Les Joujoux*, Librairie Octave, Paris, 1924

Cassell's Household Guide, volume III, Cassell, Petter & Galpin, London, c.1875

Cieslik, Jürgen and Marianne, *Puppen Bestimmungs Buch*, Marianne Cieslik Verlag, Jülich, 1984; *German Doll Encyclopedia*, Hobby House Press, Inc., Maryland, 1985

Coleman, Dorothy S., Elizabeth A. and Evelyn J., *The Collector's Encyclopedia of Dolls*, volume I, Crown Publishers, Inc., New York, 1968/Robert Hale, London, 1970; volume II, Crown Publishers, Inc., New York, 1986

Cremer, Henry, *The Toys of the Little Folk of all Ages and Countries*, London, 1873

Crump, Lucy, *Nursery Life 300 Years Ago: The Story of the Dauphin of France*, Routledge, 1929

Desmonde, Kay, *Dolls and Dolls' Houses*, Charles Letts & Co., London, 1972

Deutsches Spielwarenzeitung, Nuremberg–Bamberg, 1925

Dickens, Charles, *The Cricket on the Hearth*, Chapman & Hall Ltd, London, 1846

Earnshaw, Nora, *Collecting Dolls*, William Collins & Co., Ltd, London, 1987

Ellis, Robert (ed), *Catalogue of the Great Exhibition of 1851*, volumes I, II and III, Spicer Brothers and W. Clowes & Sons, London, 1851

Flick, Pauline, *The Dolls' House Book*, William Collins Sons & Co., Ltd, London, 1973

Gerken, Jo Elizabeth, *Wonderful Dolls of Papier Mâché*, Doll Research Associates, Lincoln, Nebraska, 1970

Greene, Vivien, *English Dolls' Houses of the 18th and 19th Centuries*, B.T. Batsford Ltd, London, 1955 (reissued 1979)

—, *Family Dolls' Houses*, G. Bell & Sons, London, 1973

Gröber, Karl, *Children's Toys of Bygone Days*, B.T. Batsford Ltd, London, 1928

Gross, Claus-Peter, *1871–1918: Verliebt—Verlobt—Verheiratet*, Willmuth Arenhövel Verlag, Berlin, 1986

Hasluck, Paul N., *Painters' Oils, Colours and Varnishes*, Cassell & Co. Ltd, London, 1910

Heal, Ambrose, *The Sign Boards of Old London Shops*, B.T. Batsford Ltd, London, 1947

Hennig, Clare and Landolt, Margie, *Alte Puppen—Neugesehen (Old Dolls in Company)*, Ineichen Verlag, Zurich, 1984

Hillier, Mary, *Dolls and Doll Makers*, Weidenfeld & Nicolson, London, 1968/G.P. Putnam's Sons, New York, 1968

—, *Automata and Mechanical Toys*, Jupiter, London, 1976

Hillier, Mary (ed) and Fawdrey, M., *Pollock's Dictionary of English Dolls*, Robert Hale, London, 1982

Hughes, G. Bernard, *Collecting Miniature Antiques*, William Heinemann, London, 1973

Jacobs, Flora Gill, *A History of Dolls' Houses*, Charles Scribner's Sons, New York, 1953/Cassell & Co. Ltd, London, 1954

—, *A Book of Dolls and Dolls' Houses*, Charles E. Tuttle & Co., Inc., Rutland, Vermont, 1967

—, *Dolls' Houses in America*, Charles Scribner's Sons, New York, 1974
—, *Victorian Dolls' Houses*, Washington Dolls' House and Toy Museum, Washington D.C., 1978
Jackson, Mrs F. Nevill, "A Tudor House" in *The Connoisseur*, 1927
Jekyll, Gertrude, *Old English Household Life*, B.T. Batsford Ltd, London, 1925
Joy, E.T., *English Furniture A.D. 43–1950*, B.T. Batsford Ltd, London, 1962
Kemp, Wilfred, *The Practical Plasterer*, London 1893 (reprinted Crosby, Lockwood & Son, 1926)
King, Constance Eileen, *The Collector's History of Dolls' Houses*, Robert Hale, London, 1983
Lines Brothers, *Looking Backwards, Looking Forwards*, 1958
McClintock, Marshall and Inez, *Toys in America*, Public Affairs Press, Washington, D.C., 1961
Marshall, Barbara and Francis, Mary Harris, *Dolls' House and Miniature Museum Book*, Toy and Miniature Museum of Kansas, Kansas City, Missouri 64112
Marwitz, Christa von der, *Das Gontard'sche Puppenhaus im Historischen Museum, Frankfurt am Main*, H. Kunz Verlag, Kelkheim, 1987
Mayhew, Henry, *Mayhew's London* (edited Peter Quennell), Pilot Press, 1949
—, *German Life and Manners as Seen in Saxony*, William Allen & Co., 1865
Merrit, Mary, "Guide Book", Mary Merrit's Doll Museum, Douglassville, Pennsylvania, 1963 (second edition 1972)
Morris, Christopher (ed), *The Journeys of Celia Fiennes*, Cresset Press, London, 1947
Ramsey, L.G.G. (editor), *The Complete Encyclopedia of Antiques*, Connoisseur Publications, London, 1962
Sala, George, "Paris Revisited" from a series in *The Daily Telegraph*, London, 1878–9
Sidgwick, Mrs Alfred, *Home Life in Germany*, Methuen, London, 1908
Stalker, J.S. and Parker, G., *Treatise of Japanning and Varnishing*, Oxford, 1688 (reprinted Alec Tiranti, London, 1971)
Whitton, Blair (ed), *Bliss Toys and Doll Houses*, Dover Publications Inc., New York, 1979
Woodford, John, *Georgian Houses for All*, Routledge & Kegan Paul Ltd, London, 1978
Yarwood, Doreen, *The Architecture of England*, B.T. Batsford Ltd, London, 1963

NEWSPAPERS AND PERIODICALS

Games and Toys, 1916, 1918 and 1922 on
Girls' Realm, 1901, 1902 and 1907
Magazines and catalogues published by Hobbies Ltd between 1934 and 1987
Illustrated London News, 1851, 1861, 1865 and 1878 ("Exposition Universelle")
Toy and Fancy Goods Trader, 1918–24
Trade journals and magazines, too numerous to mention, published during the 19th and early 20th centuries

AUCTION HOUSE CATALOGUES

Christie's, South Kensington, London
Ineichen, Zurich
Phillips, London
Sotheby's, London and Chester

RECOMMENDED SUBSCRIPTION

The International Dolls' House News
P.O. Box 79
Southampton SO9 7EZ

ACKNOWLEDGEMENTS

I should like to thank everyone who has helped in the production of this book: Norah and Cyril Bardwell; Olivia Bristol, Christie's, South Kensington (for enthusiastic assistance and support); Diane Buck, the helpful curator, and Catherine Hanlon of the Wenham Museum, Massachusetts; Graham Buckingham; Bunny Campione; Sotheby's, London, for unstinting help and staunch support; Dorothy S., Elizabeth A. and Evelyn J. Coleman; Kay Desmonde; Elizabeth M. Donoghue, former curator, Wenham Museum, Massachusetts (for invaluable advice, unfailing assistance and for loving friendship); Judy Emerson of the Margaret Woodberry Strong Museum; Pauline Flick (for support and sympathy); Rosalinde Gers (for information from Berlin); B. and H. Harden; Harrogate Processing Laboratories, incorporating Frank Newbould Photography (for magnificent help and assistance); Carl and Waltraud Heiss (for reference from Einsingen-Ulm); Mary Hillier; my dear friend Mary Johnson (for welcome support); Hilary Kay of Sotheby's, London; Paul Larkin, Curator of Collections, and Alan Garlick, Assistant Curator, The Abbey House Museum, Kirkstall (for enthusiastic help and great patience); the staff of Leeds Patent Library, especially Peter Robson and Stefania Stephenson; Kevin Matthews; Celia Mayfield; Roger Pearson; Alan Robertson; Priscilla Spurrier of Christie's; Rodney Tennent and Sally; Grace and Alfred Thompson, wonderful supportive friends; and Victor H. Watson, CBE, Chairman, John Waddington Ltd.

Grateful thanks too, to the curators of the Nationals Museet, Copenhagen; the Historisches Museum, Frankfurt-am-Main; and the Mary Merritt Doll Museum, Pennsylvania; and to the photographic librarians of Phillips, London, and of Sotheby's, Chester.

I should also like to express my gratitude to the doyenne of English dolls' houses, my dear friend and travelling companion and the curator of the Rotunda Museum of Dolls' Houses, Oxford, Vivien Greene, for many years of happy friendship and unfailing support; and to the doyenne of American dolls' houses, my dear friend Flora Gill Jacobs and her husband Ephraim. For swift assistance gratefully received I acknowledge the co-founders of the Toy and Miniature Museum of Kansas City, Barbara Marshall and Mary Harris Francis, who gave me the red carpet treatment, and their enthusiastic staff, Sandi Russell and Lambert Zachary; Mrs Elizabeth Conran, the Curator of The Bowes Museum, Barnard Castle, County Durham, and her assistant Dinah Jones; Mrs Rosemary Gillet, the Curator of Nunnington Hall, Mrs Joy Ramsay and the administrative staff of the National Trust, York, and Mr Roger Whitworth of the National Trust; Lord St Oswald of Nostell Priory; Christine Kohler of Auktionhaus Ineichen, Zurich, and Margie Landolt of Basel. My thanks also to my many friends in the Doll Club of Great Britain and the United Federation of Doll Clubs in America. Finally to my dear friend Shila Vandervalt, who helped me with her time and generosity in Kansas City, Missouri, I give heartfelt thanks and much gratitude.

Memories will always linger, and I take this opportunity to record my gratitude for the unfailing help and support and sincere friendship over many years of the late Diana Buckingham (and my two god-children Victoria and Emma Buckingham); Peggy Byford; Magda Byfield; Eleanor J. Carter; Amy Hadwen; Irene Blair Hickman; god-daughter Nora Kingsnorth; Lucy Newbould; Faith Rowntree; Mrs Alec Ison (Nonie Rowntree) and Gwen White.

INDEX

Page numbers in *italics* refer to picture captions